THE MACHINERY OF

GRAVITY

GENERALIZED EQUIVALENCE

DAVID FRANKLIN

ACKNOWLEDGEMENT

Above all others I cite my wife Loretta, without whom this book would have been impossible. Her support as a mathematician, as an editor, as an honest critic and, above all else, a willingness to put up with my apparently non- ending focus on this project (absent from the usual business of being a husband), (except for still doing the dishes) was not an easy task. I thank you my love.

To my two editors, Jonathan Sisler, who got me started on the writing with many corrections and suggestions, and Michael Zierler who took over the guidance task and finally brought the manuscript to its final form.

A special thanks to John D. Anderson who aimed me in the right direction as I tried to understand the anomalous acceleration of Pioneer 10 and 11. His taking me seriously was encouraging.

DEDICATIONS

I owe so much to so many individuals, who collectively brought my brain and knowledge to the level enabling me to write this book, that it is difficult to remember all of you. So, if I fail to mention you, take it as a failure of a less-than-perfect remembrance of my past interactions.

To Marty Siegmann, my first boss many years ago who taught me to think in his unusual, but effective, way; and likewise to Julie Adelsberg, my second boss, who taught us all to write reports, (some felt) excessively insistent to his elevated standards for simplicity and clarity.

To Nat Durlach, Charlotte Reed, Pat Zurek, Pat Peterson, Kim Kirchwey, Bill Rabinowitz, and the rest of the gang at MIT who collectively have made my life enjoyable these last thirty years.

Likewise to the staff of Audiological Engineering who made my life enjoyable (at times interesting) and convinced me I was not a good businessperson.

To Era Edell, who started as my legal advisor, but became a friend.

To Peter Mengert who gave my thinking a name and Gustav Sunderlund who was a pleasure to know and work with.

To my children, Michael, Joanne, Joseph, and their brother Paul Ernest your very existences sustain me and complete me in ways that enable me to do this work and any comments you make about it (so long as they are positive) are appreciated.

To Lydia Hughes who shaped my early life and to whom I owe a great deal. Thank you.

CONTENTS

INTRODUCTION

Einstein wrote the following in a little book published posthumously:

"---let us imagine a region of [empty] space [containing] a spacious chest resembling a room with an observer inside equipped with apparatus. ----- To the middle of the lid of the chest is fixed externally a hook and a 'being' ----begins pulling at this with a constant force. The chest together with the observer then begin to move 'upward' with a uniformly accelerated motion.----But how does the man in the chest regard the process?----if he does not wish to be laid out on the floor, he must take up the pressure with his legs. If he releases a body which he held in his hand---the body will approach the floor with an accelerated motion."

Finally the observer discovers the hook in the ceiling and attached rope and concludes he is suspended on the earth. Then Einstein writes:

"Ought we smile at the man and say he errs in his conclusion? I do not believe we ought to---we must rather admit that his mode of grasping the situation violates neither reason nor known mechanical laws."

He then expands on this imagined experiment pointing out it leads to a strong argument for extending the postulates of relativity to include bodies of reference which are accelerated with respect to one another: *"a generalized postulate of relativity."* This begins his essential thought for a theory where non-uniform motion, due to gravity, is _equated to_ "real acceleration." Eventually this results in his development of General Relativity.

Notice in particular my use of the phrase "equated to", not "replaced by". Indeed, in the same little book Einstein makes clear that one cannot in general replace any gravity field by acceleration. He writes:

"It is, for instance, impossible to choose a body of reference such that, as judged from it, the gravitation field of the earth (in its entirety) vanishes."

While his observation is true, there is a different way of interpreting this experiment that *does* enable one to *replace* gravity with acceleration, including bodies such as the earth (in its entirety). That is to make note of the inversion in his model where the force the observer experiences is due to an action of the floor *pushing upward.* This is just the opposite of our usual thinking of gravity being an attractive force. Can one show this view is internally consistent, without contradictions? Is this a possible explanation that gravity and acceleration are one and the same?

The purpose of this book is to explore that possibility and, as you will see, it may be a perfectly good explanation for the machinery underlying gravity's behavior. That is, gravity is caused by all matter expanding, endlessly, forever!

However, the path to understanding this new way of thinking about gravity encounters a number of obstacles, not the least being emotional. The idea of gravity being an attractive force has existed back into prehistoric times, affected our language and has been enshrined by Newton and more lately, by Einstein. It would be unprecedent in human behavior if this radical change did not cause consternation at the very least.

I recognize this and acknowledge that my course will encounter skepticism but believe the arguments I will present will convince a few of you, entertain more of you, and undoubtedly give many of you something to argue about. For myself it allows me to get off my chest an idea that has been resident in my brain far too long and I want to get back to other aspects of my life. Here it is, a new truth or an elaborate fantasy. Make of it what you will.

In a work like this it is inevitable that I use some equations, but since I am addressing this work to laypersons interested in physics as well as the physics community, I have minimized them as much as I can. I've kept them simple and provided non-mathematical descriptions of the meanings in all cases.

Preceding some chapters a brief set of key concepts are provided to enable those not familiar with physics be able to follow the arguments. Also at the beginning of most chapters (when I think it is useful) is a summary of the content in easily understood language that avoids all mathematics and confusing details telling what the chapter is saying.

A couple of observations: Einstein traveled this same path when he developed his own theory of gravity, General Relativity. I have asked myself why he did not see the same solution I present here? I found the answer in several of his communications published on-line (volume 4-"Writings,Berlin years, 1914-1918"). In essence, from the very beginning of his work Einstein focused on the need to consider an active role for space itself as a character in the actions of gravity. In taking this position, his gravitational theory is not just a description of the way gravity acts, but a _causa_l consequence of nature's behavior, emphasis on the word "causal".

This view later echoed his dislike of quantum theory where he once said, god does not play dice with the universe. One consequence of his development, his rejection from the outset that the universe was other than totally ordered, was that his view of gravity failed to fit into what is called the standard model of particle physics that should in principle contain all the known forces of nature, but only includes three, gravity not fitting in.

In this context I should mention the work of Erik Verlinde, a Dutch physicist, who postulates that gravity is an _emergent process_ and not one of the fundamental forces of nature, _"something that "just is" ---that gravity is actually the result of the positions of quantum bodies, similar to the way temperature is derived from the motions of individual particles."_ He adds that, _"---- Einstein's theory looked more like the laws of thermodynamics),(because of what black holes have taught us), and the laws of thermodynamics we know can be derived by thinking about the microscopic constituents that are describing matter."_

I agree with Verlinde's view and while my development for the most part agrees with General Relativity, I think GR is descriptive, not causal (in the sense of explaining what causes gravity) as Einstein believed. At bottom I think it is a statistical process just as Verlinde suggests.

If Verlinde's ideas are at the top of my list of thinkers whose thoughts I admire, then Mark McCutcheon, who had a really good idea, the expansion notion, is not. While I give him credit for his insight, I cannot praise his work. I am referring to his unfortunate book "*The Final Theory*", first published in 2002. From the title alone it should give warning to any serious student of physics. He begins the book by pulverizing Newton's gravity law, bending it to his needs, and from that point descends into a pit fashioned by his own digging.

Can McCutcheon at least claim to be the originator of the expansion idea? I'm afraid not. This applies to me as well. If evidence of that is assigned to the individual to first put it into recent print, then the award goes to Scott Adams in his wonderful book, "Dilbert Future" (copyright 1997). Even earlier than that in the 1960's, the famous physicist Paul Dirac, put forth a similar speculation. I suspect the notion has been around much earlier, but when or by whom it first emerged shall forever remain a mystery.

Finally, a few words about how this book is organized. The first several chapters explore the concept and how to interpret it. As you will see it requires a change in thinking about certain phenomena that violate our usual under-standings and shows how powerful our ideas of reality are ruled by our train-ing, beliefs, and daily experiences.

The middle part of the book applies this new concept to various prob-lems in physics where solutions are not known or simpler solutions are found using this new approach. These include the calculation of the advance of the perihelia of Mercury, Venus and the Earth, the bending of light by the Sun, the expansion of the earth, the relation between the expansion hypothesis and the Hubble constant and more.

The last part of the book critiques these teachings and discusses how they fit into our present understandings of gravity and the evolution of the cosmos. It also includes an essay on truth and the methods of science here in the West: Certainly not the only way to understand the goings-on of the universe, but an approach rooted in Greek teachings that has become our mainstream way of thinking ever since.

Dr. Peter Mengert, mathematician (unhappily now deceased) suggested the name, **GENERALIZED EQUIVALENCE (GEQ)** for the Expansion Hypothesis because of its relation to Einstein's "Principle of Equivalence" that preceded and was important in his development of General Relativity (**GR**). I will use these shortened labels in the body of this book.

What is Gravity; what is its machinery? Let's take a look.

CHAPTER I

The Machinery Of Gravity

"What is gravity? But is this such a simple law? What about the machinery of it?[What makes it go[?] Newton made no hypothesis about this... No one has since given any machinery. **RICHARD P. FEYNMAN in his lectures on physics; June 1963**

Is it possible that gravity is a pushing phenomenon caused by the earth expanding forever? Contrast this with Newton's view that gravity is an attractive force between all matter. The second law of thermodynamics, that you are familiar with every time you pour yourself a hot cup of coffee only to have it cool down more than your palate would like, shows this notion is not entirely silly. What's happening of course is that the hot cup is losing energy in the form of heat to the cooler surroundings through a combination of electromagnetic radiation and conduction.

This is not a minor consideration in understanding how the universe works. Entropy (the measure of disorder) is an inevitable consequence of the above law and, if any law of physics is universally applicable, it is the second law of thermodynamics. Entropy in the long run <u>always</u> increases: Life dies and metal rusts or otherwise disintegrates. Civilizations in the end always fail. Even the so- called "immortal jelly fish" finally succumbs to the inevitable loss of its environment, as do we all; the great "heat death of the universe" where all matter radiates its energy away and what is left is a uniform soup of equal temperature everywhere—a state of maximum disorder.

But what does this have to do with gravity? Rather a lot! Another way of stating the law, a way represented by the cooling cup of coffee, is that any time you have an intense collection of energy surrounded by lower energy, the higher energy will leak out and flow towards the lower, thus creating more disorder. Given this inevitability, what is more energetic than matter? Is not Einstein's conclusion from his 1905 paper that $E = mc^2$ of note? Taking that finding (and it is one of those understandings in physics that is settled and cannot be questioned), how is it that we continue to view *matter*—all matter, us included—as remaining essentially constant in size?

There are more clues that have to be considered as well.

First, there is our everyday experience of the similarity between gravitational force and the experience of being pushed into our seat in an accelerated car. Even more to the point are Einstein's observations that an observer in a closed accelerated chest, pulled upward by an "angel," is not able to distinguish between that state of affairs and with being at rest on the surface of the earth. This insight was of central importance for Einstein in developing his new theory of gravity (General Relativity). While Einstein's model has limited connection between the two states (only valid in a small enclosure where the field intensity is constant), it begs one to wonder if his idea can be extended to *all space*. Can one claim that gravity *is* acceleration, that they are one and the same? That is the goal of this book.

These observations should not be ignored. While I cannot state this expansion hypothesis as anything more than a conjecture at this point, it is certainly worthwhile to examine it in detail. However, there are issues that seem to contradict the idea.

For example, one of the necessary properties of this expansion notion is that <u>all</u> matter must be expanding exactly the same way. Thus, we are left in the situation of not being able to distinguish that dynamic state from our usual understanding that all matter remains essentially static in relative sizes. That is, what remains time invariant is the <u>ratio of everything</u> regardless of densities, geometries, states (including subatomic down to the quantum level), and every large thing up to and including galactic scales and even human beings. This is a rather large and forbidding set of requirements, to say the least.

One does not have to dig very deeply into what can be stated as a "truth" in physics to discover a conflict with the expansion hypothesis: that is, Newton's law of gravity (Acceleration=- GM/R2). To clarify this issue let me present a thought experiment, one of Einstein's favorite tools for analyzing a problem. It goes like this.

Consider two large spherical masses of the same size but with very different densities. According to Newton's law, as written above with a negative sign on the right-hand size of the equation, if small test masses are placed near both spheres, the test mass near the denser sphere will fall faster than the other one.

But if I want to imagine this represents the expansion idea, I have to replace the negative sign with a plus sign. The equation now says, both spheres expand and the test masses stay at rest. The problem is the mass term on the right-hand side of the equation (M) that says, the denser sphere expands more rapidly than the other so they become different sizes. This is <u>not</u> what happens in reality!

This means the expansion hypothesis is wrong or, in medical/mystery terms, DOA. Of course, the whole point of this book is to show that this is not the case, and the very first task is to deal with this. As you will see, the solution is not as difficult as one might think.

Also, there are experiments performed that suggest the expansion hypothesis might have legs; that it might be correct.

These include a series of studies, the so-called anomalous accelerations observed by NASA during the launching of the deep-space probes, Pioneer 10 and II. They have never been completely explained even though one of the scientists involved, Dr. Slava Turyshev, has claimed a solution for one of the three anomalies observed. An apparent extra force from the Sun was observed that shouldn't exist, as well as a yearly variation in that force, and a very much larger daily variation in that force. While the physics community has accepted his explanation as correct, the strength of his solution fails to reach the level of certainty usually required by the science community and fails to address the two remaining anomalous terms observed. As you will see in the body of this book a solution is offered that addresses all three anomalous terms using teachings from the expansion hypothesis.

In addition to the Pioneer anomalies, a few physicists are exploring curious data that suggest the earth has increased its size over the last 175 million years. In doing so they are outside the present paradigm physicists use and approach the issue with no guiding principles. I examine this suspicion that the earth is expanding with a new paradigm based on the teachings of this book and demonstrate that, with their somewhat flawed methodology, these physicists came close to the proper answer. I describe a more constrained experiment that supports both the expansion hypothesis and the idea the earth has indeed increased its size.

Let me start this exploration and show it is a viable, internally consistent possibility, well worth a bit of effort. If nothing else, it is an interesting way of looking at the way the universe works.

SUMMARY OF CHAPTER 2

This chapter starts by pointing out that Newton's gravity law has to be considered a "truth" and that the expansion notion has to agree with it. But the expansion notion appears to violate it for the following reason.

His law says any large mass (say, the earth) causes any small mass (call it a test mass) to accelerate towards the center of the large mass at a rate proportional to the magnitude of the large mass. If the large mass is either greater or less massy, the acceleration towards its center will accordingly be greater or less.

So, if you reverse the process and say the large mass gets larger in time, and that's where the acceleration comes from, then it must be true that the acceleration gets greater when the mass gets greater. Accordingly, if I have two large spheres of the same size, but one is massier than the other, the denser one must get larger than the lighter one and, of course, that's not what we see.

But in reality, this is not what happens. This is where Mark McCutcheon got lost and never recovered. The proper solution explained in this chapter utilizes Einstein's conclusion in Special Relativity that nothing can go faster than light speed and uses some simple calculus to find the size of any sphere after a long time has passed.

It turns out the final size is very nearly independent of its mass (density), but not exactly so. The difference is extremely small and not generally measurable, except in situations where very long times and very great distances are involved.

Since the expansion applies to all matter, including us, we cannot see differences in the sizes of any rigid bodies, independent of their densities or shapes, and we cannot tell the difference between gravity caused by an attractive force or by an expansion process. Note that this does not claim the expansion idea is really happening, but it does say it might be happening, as far as agreeing with Newtonian gravity. This chapter derives the mathematical relations for the difference term, and later chapters use the equations to solve a bunch of physics problems. I point out that since the difference term exists, it provides a means to further test the expansion notion.

KEY CONCEPTS FOR CHAPTER 2

1. **Newton's Gravity Law** Newton's gravity law says any object accelerates (falls) towards any large object (for example the earth) at a rate that increases as the object gets closer to the earth.

The formula describing this is: Acceleration $=-GM/R^2$

Read this as any small mass near a large mass M accelerates *towards* (falls towards) that large mass at a rate that decreases by the square of its distance R from the center of M. The negative sign indicates it is an attractive force. The magnitude of M determines how great the acceleration is.

If you replace the minus sign with a plus sign, the equation now reads the reverse: Any small mass near a large mass remains at rest, and the large mass M expands towards the small mass.

2. **The velocity of light is the same for all observers independent of their states of motion.**

This is very different than the way sound behaves. You can catch up to a sound wave if you move fast enough (about 350 meters per second). When an aircraft reaches that speed, it catches up with the sound of its engine and generates a sonic boom. This is one reason the French Concord stopped flying!

For light, no matter how fast you move, light is moving away from you at the same rate (about 3×10^8 meters per second). No object can move faster than the speed of light or catch up with it.

3. **No material object can reach or exceed the speed of light.**

If you try to accelerate something without limit, you will find it cannot reach light speed. When you apply forces to try to make an object go faster you will discover its speed increases less and less as it gets closer to light speed no matter how strong the force is.

CHAPTER 2

Expansion Process Vs. Newton's Theory

The point of this book is to introduce a new meaning to our understanding of how gravity behaves that is in direct conflict with our present belief. That is rather than being an attractive phenomenon, "all masses are attracted to one another", instead. "all masses grow towards one another". In doing this I introduce a new and different paradigm into our thought processes and all of our instincts and beliefs fight against this change

The first key issue lies in the apparent negative statement against the GEQ concept comes from Newton's gravity law expressed with a plus sign on the right-hand side of the equation.

$$\text{Acceleration} = +GM/R^2$$

The meaning in the GEQ interpretation of this is, " any object, say a sphere, gets larger in time in proportion to the mass (the M) so any spheres with different masses will become different sizes as they expand."

This of course is not what we see; all so-called rigid objects *do not* change sizes in time! So my first task is to show the above intuitive belief is not true.

To show this let me start by defining what we mean when we talk about the "*dimension*" of something. A useful definition is in terms of the velocity of light in a vacuum, usually symbolized by the letter "c". What is particularly useful about such a definition is that the velocity of light is the same for all observers so no matter what reference system an observer is in, how he is moving, he can always use the same definition. Furthermore, the peculiar and most useful property of light speed is that no physical object can go faster than "c".

A simple and valuable expression of this definition is:

$$\text{Length} = c*t$$

That should be read as; "the length of an object is defined by the lapsed time for a pulse of light to travel from one end of an object to the other end."

An example of this is if I measure the "time-length" of a meter stick I will find it is 13.28×10^{-9} seconds long; and if you, in a coordinate system moving at some large velocity relative to me, likewise measure a meter stick, you will get the same result. OK?

Now let me do a thought experiment according to two different points of view: First with the assumption "rigid rods" are really rigid, do not change length in time (and this of course is the standard view); the second experiment in accordance with the GEQ view that "rigid lengths" actually do change length in time.

Conventional view: Say I have a meter stick that I use to measure the radius of a sphere and find it to be 1 meter. A little time passes and I make a second such measurement and nothing has changed, the radius is as before, 1 meter.

Now let me repeat the same sequence of measurements in accordance with GEQ point of view using you: First measurement of radius, 1 meter. Second measurement a bit time later: What do you see?

That depends: If you are "in" the universe you are subject to its rules and see the same thing I do, but if you are "outside" the universe and not subject to its rules, what you see is the radius has gotten larger between the first measurement and the second! For the sake of argument let's say you see the radius has doubled so it now appears (to you) as 2 meters.

But what do you see? If you are inside the universe, subject to its rules, and what happens to the sphere, must also happen to you and the measurement stick. If you have doubled in size just has the sphere, then you will see no change in the sphere because your measurement stick still registers the same length for the radius. You cannot distinguish between a static situation (no change) and a dynamic situation (change but ratio of sizes unchanged). So, if you are "in" the universe you see the radius unchanged.

There is an important corollary to this and that is that wavelengths of light and frequencies must change in time to match the changes in physical objects and distances. Given the intimate relation between light energy and matter this requirement is a consequence of the expansion and of the observation that all observers see the velocity of light the same. I use light to measure the length of my meter stick in effect saying *"it's so many wavelengths long at some particular frequency"* so clearly it is so or there is no expansion and GEQ is dead.

On the other hand, Newton's gravity law written as the law of expansion has that nasty "M" on the right-hand side of the equation that implies objects with different densities (spheres or any other form) will expand at different rates and sizes will not be time invariant: A necessary requirement for GEQ to have legs. Looking at the GEQ expansion basis appears to assure us that this condition cannot be met.

Surprisingly, this turns out not to be true mainly because of the speed limit imposed on physical objects that they can never exceed the velocity of light.

As a metaphor consider two balloons expanding at different rates. Initially the balloon expanding faster gets larger because its radial velocity increases more rapidly, but as it nears the velocity of light, its growth rate becomes just below c and becomes nearly constant; the slower expanding balloon speeds up, catches up and from that time onward, they remain _very nearly_ the same size even as they continue to grow in dimension. According to GEQ we are in that state of dimensions since in concept everything has been expanding for a very long time.

Because of this speed limit our intuition is wrong and this can be shown by the following sequences of equations using calculus to find the size of an arbitrary sphere as it expands in accordance with Newton's gravity law, as GEQ would have it, with a plus sign. Here is that derivation for those of you comfortable in the language of calculus: For those of you not comfortable, don't worry, I will explain what is going on here and provide you with the important results.

DERIVATION OF EXPANSION EQUATION

I start by considering an arbitrary sphere that expands in accordance with Newton's gravity law with a plus sign:

$$Acceleration = \frac{d^2R}{dt^2} = +\frac{GM}{R^2} \tag{1}$$

I note: $d\left(\frac{dR}{dt}\right)^2 = 2\left(\frac{dR}{dt}\right)\left(\frac{d^2R}{dt^2}\right) dt = 2\left(\frac{d^2R}{dt^2}\right) dR = \left(-\frac{2GM}{R}\right)$ (2)

Assuming that the initial velocity is zero and (from Special Relativity) the maximum velocity reached is the velocity of light (c), I obtain:

$$Velocity = \frac{dr}{dt} = \left(C^2 - \frac{2GM}{R}\right)^{1/2} \tag{3}$$

If I define the minimum size the sphere had at the beginning of the process (t =0) as (Rmin),

I can write an additional relationship that is identical to the Schwarzschild limit.

$$Schwarzschildlimit = R_{min} = \frac{2GM}{C^2}$$
$$\tag{4}$$

Let me invert the velocity term to the form dt/dR :

$$\frac{dt}{dR} = \frac{1}{\left(C^2 - \frac{2GM}{R}\right)^{1/2}} \tag{5}$$

Integrate to obtain the relation between time (t) and size (R). Solve for the constant of integration, rearrange, and replace all terms of form (2GM/ C²) with Rmin. Properly combine terms and I finally obtain:

$$t_{eq} = \frac{R}{C}\left(1 - \frac{R_{min}}{R}\right)^{1/2} + \left(\frac{R_{min}}{C}\right) ln\left[\left(\frac{R}{R_{min}}\right)^{\frac{1}{2}} + \left(\frac{R-R_{min}}{R_{min}}\right)^{\frac{1}{2}}\right] \tag{6}$$

In equation (6), I label the left-hand time term as "teq" for convenience later in this development. After a long time passes, the first term dominates and the equation converges to:

$$t = \frac{R}{C} \quad \text{or} \quad R = C * t \tag{7}$$

Equation (6) multiplied by C tells me the _exact size_ of the sphere at any time and includes "M" buried in R$_{min}$. **Equation (7)** tells me _the approximate size of the sphere at any time_

and note, *the mass term M is missing in equation (7)!*

So look what happened. I started with the GEQ version of Newton's law, calculated what the radius would be of an <u>arbitrary</u> sphere after some time had passed, and discovered it would be *exactly* given by the mess of equation

(6), or *approximately* by the simple expression of equation (7), R =c*t, that looks the same as the definition above of length expressed as an elapsed time!

Neat result, but what does it mean in the context of GEQ?

<u>First</u>, I did this calculation for an *arbitrary* sphere, so the results would be the same for *any* sphere independent of its mass. So to the extent the sphere's size is determined by equation (7), all spheres do not change relative sizes as they expand!

<u>Second</u>, to the extent the sphere's size is determined by equation (6), there *is* a difference in the sizes of spheres of different densities. The fact is the difference is vanishingly small but can be measured under certain conditions, so GEQ can be falsified (or not) by measurement to see if the effect exists or not!

In other words, it is of the utmost importance that there is a very small difference in dimensions predicted by equations (6) versus the equation (7) prediction, that equation (7) has no M in it, and this difference can in principle be measured! The point is this provides me a way of determining by experiment if all matter really is expanding or not.

However, equation (7) has taken on a new meaning in this context, representing the size of the sphere as a function of time in the same way as the observer "outside" the universe sees it. That is, if he sees the sphere doubling in size, that is what is happening *from his point of view*. That is, the sphere's radius has gone from R to 2*R. But from our point of view we *also* have doubled in size (rule of this GEQ universe), so we continue to see the sphere unchanged in size and can claim, *what remains time invariant in GEQ are sizes to the extent size is in accordance with equation (7)!* In other words, any two spheres of different densities will remain the same relative sizes as time passes.

So, equation (7) retains its meaning in this strange ever-expanding universe governed by GEQ rules!

What all the above shows is that the expansion view of GEQ does not disagree with Newton's gravity law (no observer can distinguish between his law with a plus or minus sign) and this is a necessity since his law, within certain constraints is one of those "truths" in physics that holds even though General Relativity replaces it, providing us with a deeper understanding of gravity's behavior.

Another way of looking at this result is that to the extent what I say here is correct, I can replace the present paradigm, all rigid bodies are time invariant in size, with a new paradigm, rigid bodies in fact are not time invariant in size, they only appear that way from our point of view *and* this can be shown to be false or true under some measurement conditions. ." Note that in both GR and GEQ the definitions of length and distance are more complex than suggested here and the subtleties will be dealt with in chapter 4 on time dilation.

Let me caution you that if this new paradigm is true, it would have very dramatic effects on many aspects of physics. But at this point in what is going to be a rather lengthy analysis, it can only be considered a possibility, an interesting hypothesis.

Let me add one more example, suggested by post-doc Michelle Chalupnik (Harvard), to make clear that while I used a symmetric sphere in developing the above equations, it also applies to non-symmetric objects. For example, if I construct a random shaped 2- or 3-dimensional form from a bunch of sticks of different lengths, then to the extent the stick lengths obey equation 7, no distortion in the form will be evident, so we still see the shapes as time invariant.

SUMMARY OF CHAPTER 3

This chapter starts by considering the implications of the second law of thermodynamics. That law says that all systems, large or small, always lose energy and likewise get more disordered as time passes. With this in mind it points out that the very idea of gravity causing a group of separate hunks of matter to clump together into a single larger mass seems to violate that law because the single lump has the same amount of mass but is more ordered.

To deal with this apparent conflict, it invokes the history of the process, how the universe evolved. It is pointed out that the cloud of materials has extra kinetic energy compared to the clump and the question becomes where it got that extra energy. The conclusion reached is that in the distant past, all matter had been clumped together and something happened to break up that mass into separate pieces (work was done on it increasing its energy). This view is very similar to the so-called Big Bang idea, but with the important difference that the subsequent expansion is resident in matter itself, not in space. In this model space is a more or less passive background. It points out that this process is analogous to a stretched spring that is released wherein some of the stored energy goes into kinetic energy and some goes into heat, so the process does not violate the second law. When matter clumps according to the Newtonian view, some energy is lost as heat so the more ordered state really has less energy than before it clumped. Total usable energy in the present universe is less now than it was in the past because the expansion process is not 100% efficient, some of its energy is lost to heat.

The chapter goes on to discuss why experiencing gravity as an attractive force is understandable for our everyday belief, but it fails to consider the second law and history and leads to other contradictions. Several of them are brought into view by a thought experiment involving a man jumping off a roof, and three physicists describing what happens. The experimenter who jumps holds an accelerometer in his hand so he can report on his state of motion at any time. The "teachings" used to frame the experiment include GEQ, Newtonian, and General Relativity considerations.

Rather than attempting to describe the experiment, its conclusions, and the contradictions in the Newtonian and GR views, I suggest you read the experiment description and all the materials following it. There is no mathematics involved and the material is simple to follow. The final conclusion shows the GEQ way of looking at it is simplest and contradiction-free.

KEY CONCEPTS FOR CHAPTER 3

1. **Second Law of Thermodynamics** Entropy, a measure of disorder, always increases in all systems unless some external source of energy intervenes and somehow adds energy to the system so its entropy stays constant or decreases. Always! Examples include a hot cup of coffee, tea, soup, or what have you, each of which cools unless their surrounding environment supplies more heat energy. Originally the law came from considerations of how heat energy behaved, but eventually it came to be understood as applying to all physical processes as well. It is generally believed that the second law applies everywhere without exception, even when some process appears to violate it. When objects are packed tightly together, the ordering state has lower entropy than when the objects are scattered

2. **Energy of Mass** Einstein's famous equation, $E = mc^2$ should be read simply as; Any mass at rest to any observer has a total energy equal to mass times velocity of light squared. That's it so long as the mass is at rest relative to the observer. If a mass is moving at a velocity v relative to an observer, the energy law become $E = \dfrac{mc^2}{\left(1 - \frac{v^2}{c^2}\right)^{1/2}}$

    i.e., the mass/energy increases by the quantity of its kinetic energy.

3. **Work is done** when energy is added to any system. It is defined as Work= Force times distance displaced.

The important point is that a change in position has to occur before any energy is transferred to the object acted on. If you are the object, you "feel the force," but unless you are displaced by the force, no work is done on you. If you are standing on the earth holding a weight, no work is done, no energy is added to the mass, until you lift it up. If you hold a weight up but don't move, your muscles quiver and give off heat. Whenever work is done on you, you feel force acting on you. If you feel no force acting on you even though you are moving, no energy is being added to your state of existence.

CHAPTER 3

Thermodynamics And Gravity

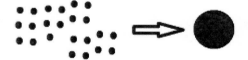

Consider the process indicated in Figure 1:

1a   1b

ENTROPY (apparently) DECREASING IN TIME

FIGURE 1

What is indicated is a cloud of small particles attracted to one another eventually coalescing into a compact sphere as indicated in Figure (1b).

Since the matter shown in Figure (1b) is the same quantity as in Figure (1a), the mass energy in both images are identical, but the "energy density" in Figure (1b) is much greater than in Figure (1a).

This model appears to be in direct contradiction to the second law. The coming together of mass from lower energy density to higher energy density is an exact analogue of heat energy flowing from a cooler surrounding to a hotter location. It should not be happening! Something is wrong, missing from the model.

What is missing is how the cloud of particles became a cloud before they coalesced into the sphere. To satisfy the demands of the second law, I have to look at the history of the process.

In fact, the energy in Figure (1a) is greater than in Figure (1b) because it has extra energy given by the cloud's kinetic energy *relative to the sphere, which I can assume is "at rest."* How did it get the extra energy? Clearly at some point an action took place, forming the cloud from a single chunk of matter, something much like the sphere shown in Figure (1b). This meets the constraints imposed by the second law; work had to have been done on the materials making up the cloud, giving it the extra energy.

When the cloud coalesces into the sphere, some of the kinetic energy is available to do work and some is lost as heat as the cloud materials impact one another and deform.

It is the same model as a spring stretched (analogous to Figure (1a)) and then allowed to collapse (analogous to Figure (1b)). In the case of the spring, some of the extra energy is available to do work, but some is lost as heat (because the stretching heats the spring, which radiates away). The net result being that the total entropy increased.

Okay, that solves that problem for this simple scenario. But what about the general case of all matter in the universe attracting all other matter in the universe?

The implication from this is all attractive gravitational forces are consequences of some historical work done on them *in the most general sense.*

Gravitational attraction is *derivative* from prior work done on it, causing mass to undergo an expansion!

If you read the introduction, you recognize these words as Erik Verlinde's comments that the source of gravity is not one of the fundamental forces of nature, "*something that 'just is,'…that gravity is actually the result of the positions of quantum bodies, similar to the way temperature is derived from the motions of individual particles*[.]" In other words, just as the temperature of a gas is due to the motion of individual particles making up the gas, so is gravity due to motion causing matter to expand, the same way a cloud of gas expands when it is heated.

While he is referring to GR, not Newtonian gravity, as I am here, it is clear that his notions apply to the earlier theory as well. From this, one should suspect that the expansion theory is more in tune with the way the universe behaves than attraction.

This suggests that some process, like a Big Bang, started the universe running. The overall machinery of the universe is expansion, not attraction.

The attraction idea is convenient as a "local" explanation of gravity, but it fails to consider the history of the process. Viewing gravity as an expansion process gets rid of the middleman and focuses attention on the underlying process: how matter is behaving.

Expressing gravity as an attractive force results in a contradiction as shown in the following thought experiment.

Imagine you are standing on the top of a tall building and decide to test Newton's Law. You jump off, free-fall to the ground, and land safely because your fellow experimenters below have arranged for a net to prevent your being squashed on impact.

The question is: What did you experience and what did the folks on the ground see? Further, what explains all aspects of the experiment <u>without any logical conflicts</u>?

Three PhD's in physics from, respectively, MIT, Princeton University (Einstein's affiliation while in the USA), and UCLA—all reputable institutions— watch you fall from their positions on the ground. They huddle briefly and announce they saw you initially held "at rest" on top of the building, accelerated downward, and finally came to rest on the ground. They add that while you were falling, they were measuring your velocity relative to their location on the ground and it clearly showed you were indeed accelerating in accordance with Newton's Law, (i.e., Acceleration = -GM/R2).

Then with a degree of disdain, both because you chose to jump, but even more because they knew you held onto the notion of expansion (they thought you were nuts, even if a bit entertaining), they asked you what you thought just happened.

You replied, "Initially I felt a force holding me in place on top of the building. When I jumped, I felt no forces on me. When I landed, I felt a force holding me up. Furthermore, throughout the experiment I was holding an accelerometer and it said I was accelerating before I jumped and after I landed, but while I was falling it registered nothing, as though I were 'at rest.'

I claim Newton's gravity law, which you just quoted, describes better what happened if I write it with a plus sign (acceleration =+GM/R2). The accelerometer readings agree with this, instead of saying I was 'at rest' when I felt a force on me. After all, Newton's second law of motion (Force = mass x acceleration) says there is motion when a force is present. What do you say to that?"

The three physicists huddle together again for a few moments, look at you with a bit more respect and finally say, "Well yes, in terms of Einstein's idea about gravity, when you are in free fall, or for that matter, in orbit around say a planet, you are in an 'inertial system' and are not accelerating, General Relativity (GR) says 'you are following your natural trajectory with no forces acting on you.' GR gets rid of the idea of a 'gravity field' and replaces it with the idea of 'curved space-time.'"

You reply, "That's good so far, but you said your measurements of my velocity increased as I fell, so something is accelerating, and if it's not me in free fall, then it must be the earth. Isn't that so? That's the only other possibility. Furthermore, when I'm either on the top of the building, or on the ground my accelerometer says I'm accelerating. How does Einstein deal with that?"

The three of them look at one another, then two bow slightly towards the one from Princeton (who happens to be a specialist in GR) and let him take over. He says, "You really don't want to go there, it gets into his Equivalence Principle *that says gravity and acceleration are much the same thing. But anyway, you have a much bigger problem with your idea. That is, on the one hand you say, quite correctly, that for your notion of expansion to work, everything independent of density must <u>expand at the same rate. So sizes of everything remain time invariant!</u> On the other hand, you claim Newton's gravity law written with a plus sign must represent the expansion. Yes?" You nod your head in agreement.*

He continues, "Well, isn't the degree of acceleration in that law different for masses of different magnitudes causing the field: More massy attractors will cause greater acceleration than less massy ones? So, dear friend, your speculations are interesting but wrong! Different objects of larger or smaller masses will quickly become different sizes. Your whole idea falls apart!" This last statement is spoken with much force and satisfaction as though this was a formal debate and you were defeated.

Your response delivered calmly but with a slightly smug smile is, "No sir, **you** *are wrong! The relative sizes of all things remain virtually constant to such a degree that one cannot measure any deviations except under some very special conditions!"*

Let me pause here and point out that the conclusions derived in Chapter 2 answer this issue neatly, showing that in fact spheres with different densities do not change in relative size.

I will leave the Princeton expert with his objection intact (for the moment) and deal with him later. For now I am more interested in demonstrating how the conventional view of gravity runs into some logical contradictions.

First, when I am standing on the top of the tall building, I feel a force acting on me, but so far as either Newtonian or GR physics are concerned, I have no motion (displacement of my location). The conventional view is *no work is being done on me because work, an increase in my energy, only occurs when physical displacement is also present.* But the accelerometer I'm holding (and its inner mechanism) says we are *moving; our kinetic energies are increasing! <u>This conflicts with the conventional definition of work!</u>*

So work is being done even though from our point of view it is not useful work that we can harness. But from the universe's point of view (or an observer outside the universe looking in), indeed work is being done. Since we get a bit crushed by the force, we heat up and radiate energy. It is not 100% efficient, and the entropy of the universe increases, as it should.

Second, when I jump off the roof, the accelerometer and my own senses indicate I am not moving, but the observers on the ground are measuring that I am falling, accelerating downward. Again, a violation of the definition of work since I feel no force and neither does the accelerometer. Furthermore, Newtonian physics specifically states that I am accelerating, but it also defines "acceleration" as the consequence of an applied force. My senses and the accelerometer attest no such force is present! There is a _contradiction between the two laws._

GR has a different message about this situation than Newtonian physics. It declares I am simply following my natural "geodesic" in spacetime, but this does little to clarify where the acceleration is coming from. Certainly, _something is_ accelerating, no matter what theory you ascribe to!

Third, the latter issue appears solved when I land on the earth and discover both my own sensibilities and my accelerometer indicate I am once again accelerating and feeling an upward force.

When the observers on the ground claimed their instrumentation indicated I was accelerating, they misinterpreted the data. What it actually indicated was they were accelerating _relative_ to me.

What all of this says to me (and I hope to you) is that paradigms are powerful forces in determining what we see and don't see! Even when our training tells us to carefully examine evidence to discover "truths," the paradigm operating in us, often subconsciously, below our conscious awareness, easily leads us to ignore the obvious and arrive at unsupportable beliefs.

As for that PhD from Princeton, the expert on GR, I can only comment that greater learning, as admirable as it is, can easily be in error when the fundamentals of his training incorporate paradigms that limit his curiosity and his ability to examine fundamental beliefs. He might have raised the GR mathematical definition of *rapidity* to explain change in speed, but he was so focused on denying the GEQ point of view, either he forgot what he had been taught, or was intrigued by the more physical explanation even though it discomfited him.

Finally, for all readers of this material who have access to the internet, I recommend that they view the interesting video of Brian Cox visiting the world's largest vacuum chamber at (*Brian Cox visits the world's biggest vacuum)* He demonstrates that a feather and a bowling ball dropped in a vacuum fall at the same rate. What is most telling in relation to this discussion is his conclusion. *The reason they stay together is because they are not falling.*

In other words, what we see (the background staying the same size), is an illusion. It only stays *constant in size* relative to us as the viewers. What is really happening ***sizes, relative to us, are time invariant. The acceleration is from the room!*** Think about it. It's a brain twister!

It will help if you imagine yourself outside the universe looking at what's going on when the feather and bowling ball start to "fall."

SUMMARY OF CHAPTER 4

There are two kinds of *time dilation* in Relativity Theory, one from SR and one from GR. Time dilation means that an observer in a given coordinate system sees time intervals differently than an observer in some other coordinate system. For SR, where the two coordinate systems are moving at very different constant velocities, the effect is symmetric in the sense that either observer sees clocks in the faster moving coordinate system as running slower than his clock.

In the version from GR where the difference between coordinate systems is intensity of gravity field, the effect is not symmetric, the observer in the less intense field sees clocks in the more intense field as running slower than his

Both of these have been tested and shown be true effects; the constant velocity one from SR by comparing two identical atomic clocks, one taken on a fast trip around the earth, the other no trip, and the travelled clock was found to have lost time. For the GR effect, measurements made of red shift between locations at different heights (gravity fields with different intensities, a good technique for time measurements) showed that effect is real as well. . Another example of this is individuals in valleys on the earth (more intense gravity field) age more slowly than mountain dwelling folk (less intense gravity field). However the effect in this case is vanishingly small, so don't bother moving.

GEQ adds a third time dilation expression from the difference between equations (6) and (7) that addresses an issue not considered in either Newtonian physics or GR; that is the growth of supposedly ridged objects saying they in fact are not "rigid" but increase in size over time. In addition, the way it is used is very different from the way the SR and GR versions are used; Specifically that it is applied to the observer's own coordinate system.

A standard definition of length for any observer in his own coordinate system is ; where c is the speed of light and delta t is the lapsed time for a pulse of light to travel end to end of some object or distance. It serves as a good equation for describing how any observer defines length in his own coordinate system. However, from the relativistic point of view it *does not* describe lengths as they are perceived in coordinate systems other than the observer's unless other considerations are applied to the equations.

The new GEQ time dilation term has two properties the other's lack; **Firstly,** it is applied to the coordinate system the observer is in; **Secondly**, it stores any changes in length over time so an experimenter can measure the sum total of length change. This is in contrast to the other time dilation terms where both are only applied to coordinate systems other than the observer's, and do not sum and store any observed changes in length.

Time dilation issues become of prime importance in later chapters of this book so the purpose of this chapter serves to introduce a reader to them, provide simple physical models demonstrating the dynamics of how they arise but makes no attempt to derive them.

KEY CONCEPTS FOR CHAPTER 4.

Coordinate Systems- as used in chapter 4, relates to states of motion. If for example I say, "my coordinate system is at rest" then any measurements I talk about are "relative to me *at rest*" unless specifically stated as otherwise. If at the same time I state, "your coordinate system is moving at a constant velocity relative to me," I have a choice in describing the motion of some object I'm observing. might say, the object has a velocity of 10 meters/ second in *my* coordinate system, **or** some other velocity in *your* coordinate system, assuredly **not** 10 meters/second, because you are moving relative to me. It would be reasonable for you to ask me; *why all this fuss about coordinate systems, why can't I just add or subtract your motion to get the answer in my coordinate system?* The answer is <u>*not always!*</u> Since the speed of light is constant for all observers, if your coordinate system is moving very fast relative to mine, I see time intervals in your system longer than in mine. This *"time dilation effect"* shows what you and I see can differ in important ways.

Time Dilation- There are two situations in physics where time intervals, as seen by any observer, lengthen compared to what an observer sees "locally" in his own coordinate system. The first relates to coordinate systems moving rapidly relative to the observer; the second when each observer is in a gravity field of different intensity.

Einstein predicted the velocity effect in his 1905 paper often referred to as "Special Relativity" (SR). It uses the "Lorentz Transformation", after Hendrik Lorentz, who derived the equations that Einstein used. The essence of the equations is that clocks run faster in a slow-moving system than they do in a fast-moving system. The effect is symmetric in that observers in each coordinate system can claim the other one is going faster than he is. If you consider the "history" of how one coordinate system moves more rapidly than the other, it becomes clear what the concept really is saying is, "clocks in higher energy systems run slower than in less energetic systems."

The gravity field time dilation term was predicted by Einstein in his 1915 paper on General Relativity. In this one, time intervals are longer (clocks run slower) for an observer in a strong gravity field than for an observer in a weaker field. The transformation uses the Schwarzschild metric that you encountered in the derivation of the expansion equations in chapter 2.

CHAPTER 4

Time Dilation And Expansion Theory

There are several equations in standard physics showing that time intervals measured by observers in different coordinate systems do not get the same results. This is at odds with the teachings in Newtonian Physics where the underlying assumption is that time flows at a constant rate for all observers independent of their states of motion or locations. Einstein showed this could not be true when he published his Special Theory of Relativity (SR) in 1905. That theory made clear that since the velocity of light is the same for all observers independent of their states of motion or locations, Newton's view had to be wrong. Observers not at rest near one another must use what are called the Lorentz transformations to compare time intervals. While this is certainly neither an obvious nor intuitive result, it has been tested many times for different scenarios and is certainly "true" in the sense that scientists use that term.

The equations developed in chapter 2 introduced a new time dilation term (difference between equation 6 and 7). Understanding how this new term fits into the standard group of transformations is central to grasping exactly what GEQ is adding to our understandings of how time and associated parameters (length and velocity in particular) behave in various settings.

I am not going to provide derivations of the following definitions, but I will provide a bit of how each definition is used and how they have been verified by experiment.

Let me start by giving an example of how this unexpected result works for the case of two observers moving at different speeds relative to each other.

Thought Experiment 1

Consider two observers, both in inertial states (no forces on either), observer B moving at half the velocity of light relative to observer A

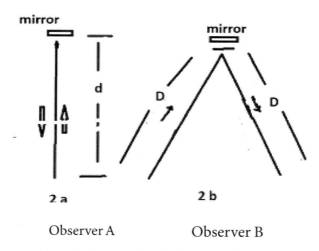

Observer A   Observer B

Time Dilation Effect (velocity difference)

FIGURE 2

Observer A emits a light pulse upward as shown towards a mirror that reflects it back to the source as is shown in Figure 2a.

The elapsed time delta is t = c*2*d.

Observer B, since he is moving at ½ the velocity of light, sees that the mirror has moved some distance to the right (as shown in Figure 2b). The pulse source has likewise moved to the right, so the path length for him is delta t' = c*2*D. Hence, he sees the time interval longer than observer A.

The equation used to describe this result is the "Lorentz Time Dilation."

formula:
$$\Delta t_B = \frac{\Delta t_A}{(1-\frac{V^2}{C^2})^{1/2}} \tag{8}$$

Notice the following:

This is a symmetric effect! Observer A would see the same result if Observer B emitted a light pulse in his coordinate system. Hence, I can interchange the A/B markers without changing the meaning of this measurement. If I slow down coordinate system B to match the velocity of observer A, the lengths d and D would now be equal, so the time intervals would be equal.

BUT, if I replace the mirror arrangement in each coordinate system with two identical clocks, wait a little while, and slow down observer B to match observer A's velocity, I will find the clock in the faster moving coordinate system has run slower (lost time) compared to the clock in the slower coordinate system. The result is not symmetric. Why is this so?

*It is because of the history of the experiment. At some time in the past, before the experiment started, coordinate system B had energy added to it (it was accelerated). The clocks in both systems can be described as dynamic "events" and rates of dynamic events are affected by the relative energy levels of otherwise identical systems. Make special note of this, as it plays a central role in what I will be commenting on shortly: **rates of dynamic events are affected by the relative energy levels of otherwise identical systems. Note in particular that equation 8 ignores the history of how observer B got to have greater velocity than observer A.***

There have been many tests showing that clocks behave as predicted concerning timing changes due to differences in velocity and gravitational fields. One of the more amusing tests occurred in October 1971. Two American physicists (Hafele and Keating) used atomic clocks aboard commercial airliners to measure time and compared the results to those of an identical clock left "at rest" in their laboratory. The results matched the predictions to very close expectations. Other tests include the corrections required for all space clocks to function properly for navigation purposes, the results of the failed Michelson-Morley light velocity measurements and too many others to list. In the case of the Michelson-Morley experiment, the Lorentz equation takes the form such that the length of a measuring stick shortens in the direction of motion from the viewpoint of an observer in a coordinate system at a lower speed.

For the purposes of how the GEQ time dilation term (difference between equation 6 and 7) fits into the panoply of other similar equations, let me move on to a discussion of how different intensity gravity fields affect measurements made by observers in differing coordinate systems. To that end, consider this new thought experiment.

(Note that in this experiment I express the time terms as t instead of so the expressions are the same as equation 6 and 7. This does not change the interpretation.)

Thought Experiment 2

Imagine two observers, observer A on the earth, the other, observer B, on the top of a very tall tower adjacent to observer A. Unlike the observers in experiment 1, both are in gravity fields and have forces acting on them. Accordingly, they are not in inertial frames. The situation is imaged in Figure 3.

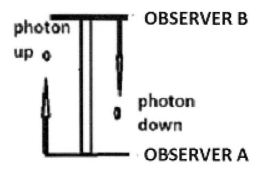

Time Dilation Effect (gravity field difference)

FIGURE 3

Observer B shines a light downward to observer A. The photons gain energy as they fall through the intervening gravity field. So, observer A sees the frequency of the light shift to blue (blue shift). Conversely, observer A shines a light up to observer B through the intervening gravity field. Photons lose energy as they climb up and shift frequency to red (red shift). Unlike the energy difference for experiment 1, the difference here is not symmetric. Clocks at the bottom of the tower rum slower than clocks at the top of the tower.

Rates of dynamic events are affected by the relative energy levels of otherwise identical systems.

The equation describing this time dilation affect is also a transformation similar to the Lorentz transformation, but uses the Schwarzschild metric and takes the form:

$$t_A = t_B (1 - \frac{2GM}{RC^2})^{1/2} \tag{9}$$

Where observer A is deep in the gravity field (say on the earth) and if observer B is imagined to be very far away from the earth B's clock would tick at 1 second per second.

Given these definitions, the equation is read as:

Observer A on the earth's surface sees his own clock tick slower than 1 second per second, and he sees observer B's time on top of the tower likewise tick slower than 1 second per second, but faster than his clock.

We know this theoretical result is correct because many carefully controlled measurements have verified the "truth" of the equation. The first of those measurements took place at Harvard University in 1959 by Robert Pound and Glenn Rebka.

However, now consider the time dilation term from GEQ, (equation 6) where I identify t_{eq} as , and R/C as .

$$_{eq} = \frac{R}{c}(1 - \frac{R_{miin}}{R})^{1/2} + \left(\frac{R_{miin}}{c}\right) ln \left[\left(\frac{R}{R_{min}}\right)^{\frac{1}{2}} + \left(\frac{R-R_{min}}{R_{min}}\right)^{\frac{1}{2}}\right] \tag{6}$$

But $\frac{R}{c}(1 - \frac{R_{miin}}{R})^{1/2} = t(1 - \frac{2GM}{Rc^2})^{1/2}$ so except for the added logarithmic terms, equations 6 nd 9 are identical.

This is a troublesome equation in the context of gravitation effects as measured by Pound/Rebka and others. We know equation (9) is correct, but equation (6) says something different because of the additional logarithmic terms.

In order to understand what's going on here remember that equation 6 arose from considering how size changed due to the GEQ assumption that all matter expanded. In view of this let me rewrite equation 6 in a form that addresses "size" instead of "time". Noting that length and time are related by "length =c*t", I can rewrite equation 6 as:

$$R_{eq} = R(1 - \frac{2GM}{Rc^2})^{1/2} + (R_{min})ln \left[\left(\frac{R}{R_{min}}\right)^{\frac{1}{2}} + \left(\frac{R-R_{min}}{R_{min}}\right)^{\frac{1}{2}}\right] \tag{10}$$

So from equation 10, I can consider how interpret the effects of this parameter and how it interacts with equation 9.

First of all, note that the addition of the logarithmic terms, causes the ratio R_{eq}/R to become greater than 1, where previously t_{eq}/t was less than 1. This means that as time passes the radius length is increasing, not decreasing.

Secondly, there is nothing in the development of equation 6 (or 10) implying it should be applied to an observer's view of lengths in some coordinate system other than his own, so this represents a very different application of equation 10 than equations 8 and 9 where both refer to coordinate systems either moving relative to the observer or resident in gravity fields of different intensities.

Thirdly, there is an implication in the development of equation 10 that the expansion process is a real dynamic property of all matter so, just like time dilation effects that are "remembered" when clocks from different coordinate systems are brought together, (differences are seen between clock times), so too here, time dependent length changes is observable by comparing earlier lengths with later lengths. This of course is the essential claim for GEQ and the validity of this claim is the main goal of this book.

This brings the issue to considering how equation 10 interacts with equation 9. Since it has to be assumed the two equations operate at the same time during any measurement event, the determinant is whether or not the slight change in distance implied by equation 10 results in any significant change in the experimental results. For this present discussion, the maximum distance for measuring gravitation time dilation was on the order of 106 meters in 1976 using a hydrogen maser mounted on a rocket. The value obtained had an accuracy of 0.007%. If you calculate the difference between equations (6) and (10) for that distance between receiver and transmitter, you find they agree to greater than 7 places ($3.3364095 \times 10\text{-}3$ sec vs. $3.3564098 \times 10\text{-}3$ sec) so the error budget in the measurement could not discern the difference since agreement in these values is on the order of 0.00001%.

On the other hand, if one searches the literature for a prior appearance of equation (6), one finds it as a description of light-time flight for electromagnetic signals passing through regions of space where planets and other bodies exist. Hence, what is clear is the appearance of equation (6) is not something new in physics, but its interpretation in GEQ is novel, as discussed below.

Focusing on the new interpretation, it is clear it suggests that any measurement of length made using a physical object, say a meter stick, is growing in length as a function of time. Any physical objects are as well! While the magnitude of this change is vanishingly small, in concept at least, one can imagine a controlled experiment to validate or negate the effect and consequently validate or negate the GEQ hypothesis. An experiment appears in a later chapter involving the growth of the earth's size. While some interest for this as an explanation for certain geological processes has been discussed and tested by various groups over the past ten years or so, no theoretical basis for the phenomenon has guided and supported the work and provided a set of criteria with which to evaluate results. GEQ can provide such criteria.

In summary, this chapter provides guidance in understanding exactly what GEQ is saying and what its verification would add to how the universe works. GEQ claims all matter is expanding in accordance with equation 6 and 7.

If the expansion <u>were exactly</u> in accordance with equation 7, we could not measure it. Since the expansion instead is <u>exactly</u> in accordance with equation 6, we can measure it under certain conditions.

GEQ <u>does not</u> include time dilation from velocity differences, nor does it include time dilation from GR, so it is an incomplete theory of gravitational behavior. It does include Einstein's teachings about the equivalence between mass and energy in as much it assumes the velocity of light is a constant for all observers.

SUMMARY OF CHAPTER 5

This chapter and the next one asks the reader to examine a number of common physical processes using the GEQ point of view. Some of the examples can be a bit difficult to accept because of the way interactions among separated bodies are described. Several violate our sense of reality, even though logic shows the descriptions are reasonable.

The main problem is that in GEQ the acceleration of the observer is taken as a primary factor in what he sees as opposed to what the object he is looking at is actually doing. For example, the object may be "at rest" while the observer is accelerating, but so far as the observer is concerned, it is the object that is accelerating. The first example in the chapter uses this kind of model to make the issue clearer. It is later examples in this chapter and in the next one that can and probably will cause mental confusions.

Let me begin by discussing the different ways Newtonian, GR, and GEQ physics deal with the concept of gravitational force.

In the Newtonian view, gravitational forces exist. Period. They are the reason masses attract one another, the reason objects can move in other than straight paths, the reason planets orbit. If forces did not exist, objects at rest would have to remain at rest, and objects in motion could never change speed or direction. Any and all interactions between discrete objects are mediated by forces of one kind or another. While space exists for Newton physics it serves as an "absolute at rest" reference system and all motions caused by forces take place in that frame with quantitative magnitudes measurable with a "clock" (time) that is the same for all observers. One consequence of the Newtonian view is that "force at a distance" occurs. Newton was never really happy with this concept but never resolved it.

In GR, once Einstein got beyond SR, he eliminated the notion of gravitation force by saying that mass caused space to curve. In the GR universe all actions are governed by geometry and force is no longer required. Planets orbit following their "natural paths," objects speed up and slow down because of the rotation of a velocity vector, and so on. To the extent mass exists, there can be no real "at rest" state because the interaction of bodies always exists and the relative motions of *all* bodies, big or small, follow their natural paths.

He explicitly uses the idea of different coordinate systems as reference frames for motions, using what are termed Lorentz transformations to allow observers in different coordinate systems to calculate what they will see in each other's systems. His space is an active element in all motions and time varies for different observers based on their locations relative to one another and their states of motion. Einstein's universe is driven by dynamic geometry and it is fairly labeled as a geometric theory. No material object can reach the velocity of light, but space can expand at superluminal velocities.

GEQ theory borrows from both Newtonian and GR physics. While GR gets rid of gravitation forces by introducing curved spacetime, GEQ gets rid of it by claiming all matter expands. Hence in GEQ, curvature of spacetime is an illusion. It then follows that GEQ does not view space as a reference system, but as a background framing that is passive. Same as GR, GEQ uses the Lorentz transformations to describe the relation among observers in different coordinate systems. GEQ does not allow anything to reach the speed of light, including space, which it views as passive.

GEQ runs into a problem if you try to view it as an expansion process because if you view it that way, you run into infinities of sizes and we don't know how to deal with infinities, so it uses the following dodge.

It notes that since, from our point of view, all rigid bodies are time invariant, we can view the universe as "Newtonian," but with several notable differences.

1. No central forces: instead the observer is accelerating.
2. When an observer is accelerated, he sees moving objects trace out curved paths.
3. For large velocities and/or large distances between the observer and the observed, you have to consider the difference between equations (6) and (7), which turns out to be a new transform closely related to the Lorentz transform for gravity field differences.

An implication of this reversal is that it now appears to the observer that space is no longer passive but possesses an "inward flow." This becomes very important in cosmology and brings the GEQ view into a kind of agreement with GR, although in a manner that is not exact. I will address this in chapters 14 and 15.

The last part of this chapter deals with the thinking process required to deal with physical problems using the GEQ point of view. There is no mathematics in this chapter; it deals only with qualitative issues.

KEY CONCEPTS FOR CHAPTER 5

Coordinate Systems- as used in this chapters, relate to states of motion. If for example I say, "my coordinate system is at rest" then any measurements I talk about are "relative to me *at rest*" unless specifically stated as otherwise. If at the same time I state, "your coordinate system is moving at a constant velocity relative to me," I have a choice in describing the motion of some object I'm observing. I might say, the object has a velocity of 10 meters/ second in *my* coordinate system, **or** some other velocity in *your* coordinate system, assuredly **not** 10 meters/second, because you are moving relative to me.

It would be reasonable for you to ask me; *why all this fuss about coordinate systems, why can't I just add or subtract your motion to get the answer in my coordinate system?* The answer is *not always!* Since the speed of light is constant for all observers, if your coordinate system is moving very fast relative to mine, I see time intervals in your system longer than in mine. This *"time dilation effect"* shows what you and I see can differ in important ways.

Time Dilation- There are two situations in physics where time intervals, as seen by any observer, lengthen compared to what an observer sees "locally" in his own coordinate system. The first relates to coordinate systems moving relative to each other. The second refers to observers in different intensity gravity fields.

Einstein predicted the velocity effect in his 1905 paper often referred to as "Special Relativity" (SR). It uses the "Lorentz Transformation", after Hendrik Lorentz, who derived the equations that Einstein used. The essence of the equations is that clocks run faster in a slow-moving system than they do in a fast-moving system. The effect is symmetric in that observers in each coordinate system can claim the other one is going faster than he is. If you consider the "history" of how one coordinate system moves more rapidly than the other, it becomes clear what the concept really is saying is, "clocks in higher energy systems run slower than in less energetic systems."

The gravity field time dilation term was predicted by Einstein in his 1915 paper on General Relativity. In this one, time intervals are longer (clocks run slower) for an observer in a strong gravity field than for an observer in a weaker field. The transformation (equation) is the Schwarzschild metric that you have encountered previously.

Chapter added a third time dilation term as noted in chapter 4. None of these time dilation plays direct roles in the following development but are mentioned here for completeness.

CHAPTER 5

Coordinate Systems And Perception

The motion of our coordinate system affects what we perceive. This is one of the most important findings of GEQ. While it is well known in mainstream physics, GEQ elevates it to a primary position as part of the new paradigm, "all matter is expanding". Here is a thought experiment demonstrating how it works:

Consider three observers watching an object moving through empty space.

The first observer, "at rest," will report the path followed by an object moving at a constant speed, left to right, in a straight path. See Figure 4.

OBSERVER AT REST

FIGURE 4

He sees the object moving with uniform motion parallel to the bottom of the page.

The second observer is moving at a constant velocity towards the object. He reports the path is not left to right but slanted at an angle. See Figure 5:

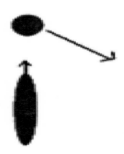

OBSERVER MOVING TOWARDS OBJECT

FIGURE 5

If you are a frequent airline traveler, you probably have seen this effect. If you look out the window and see a small plane traveling below you, it appears to move not along its axis but rather, diagonally. See Figure 6.

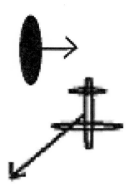

VIEW OF SMALL PLANE BELOW OBSERVER

FIGURE 6

More dramatically, if a third observer is, like the second observer, moving towards the motion but accelerating, he will report the path is a curve. Figure 7:

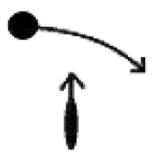

ACCELERATED OBSERVER

FIGURE 7

In Newtonian physics a force has to act on any moving object to "deflect it" from a straight path. There is no other possibility.

But GR and GEQ deal with curved paths in different ways. GR says the presence of matter causes "space to curve". An object follows its "*natural path*". GEQ says something a bit different. If an *observer is accelerated , he will see an object move in a curve*. In both approaches Newton's notion of "force on the object" is not required to support curved motion.

While both GR and GEQ agree on the results (more or less), they do not agree on what is primary. GR claims "spatial curvature" really exists (more properly called "space-time curvature"). GEQ claims spatial curvature only appears to exist because everything is getting larger (accelerating in size).

While the effect of an observer accelerating is nothing new, no observer *in* the universe can experience matter expanding and "understands" that a "rigid body" by definition is time invariant in size. His paradigm tells him that is the case. He cannot imagine that the acceleration he feels means he is being spatially displaced (work is being done on him) and must look elsewhere to explain how gravity works.

Einstein's thinking was also bound by that paradigm. At the same time he rejected Newton's idea of *"action at a distance"* (force between masses not touching), that is inherent and essential to Newton's gravity theory. In view of this and his related desire to include action *by* space as a part of the elaborate dance of gravity, his assumption that, 'mass caused space to curve" and "space caused mass to move" was, and remains, a reasonable geometric explanation for that dance.

GEQ has no argument with Einstein's approach except to claim his methods are not basic, not "true" in the sense "space really is curved" but rather "appear to us as curved". His methods, while very effective, fail to highlight the all-important consideration of "relative to an observer".

Now I would like to develop a model for GEQ that relates it to GR and Newtonian gravity, setting the stage for applying GEQ to some physical events using neither forces nor GR spatial curvature.

I started chapter 1 focusing on that pesky "M", but in getting rid of it I ended up by introducing infinity for sizes of all physical objects. While I stated that sizes are relative to us, that since we also are expanding in the same way as all other material objects, we do not have to think about "infinity" at all. Let me expand on that claim and provide a suitable model for our thought processes.

Imagine you have a meter stick and use it to measure the dimensions of the rectangle shown in Figure 8a. Your measurement says that the 2-dimensional object is 2 x 4 meters. After some time has elapsed, the time-dependent expansion has occurred a la GEQ. You measure it again as in Figure 8b.

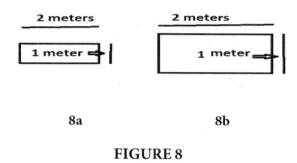

8a   8b

FIGURE 8

Since the meter stick grows in the same way as the rectangle, you report that measurements in Figure 8a results in 2 x 4 meters and measurements in Figure 8b give *identical* results! Why?

Because the equation *length = c*t is independent of an observer's coordinate system for any local measurement.*

What Figure 8b shows is what an observer <u>outside the universe</u> sees; an observer <u>inside the universe</u> sees Figure 8a, no change.

In this particular situation the different coordinate system is based on a time difference resulting in a change of scale, but you cannot see that change because the the expansion applies to you as well as the meter stick and the rectangle.)

While the example given here is just for two dimensions, it holds for three spatial dimensions so long as that little equation, *length = c*t, holds.*

It would be very useful to have a physical representation for GEQ, similar to the kind of model used in GR showing the way space behaves, forming a kind of bowl around any mass. Here is that well-known model. Sometimes it is described as a rubber-sheet effect, because any mass in the vicinity of the warping will roll towards the central mass or, as indicated in Figure 9, a second small mass orbits the central mass if it has just enough kinetic energy (velocity) to counter the rolling-inward effect.

Rubber Sheet Model

From American Association of Physics Teachers

FIGURE 9

Toward that end let me introduce two diagrams of how "mass" expands into surrounding space, (Figure 10a), and an alternative diagram, (Figure 10b), that gets rid of the "infinity" introduced by the GEQ concept and replaces the expansion, by an imagined "flow-of-space" inward expansion by an imagined "flow-of-space" inward.

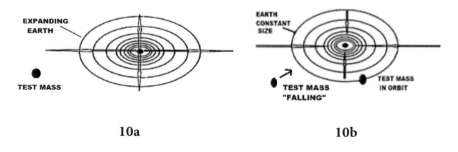

10a   10b

FIGURE 10

In Figure 10a we see a 2-dimensional representation of the earth with outward facing arrows indicating its expansion in time. Located to the lower left of the earth is a small test mass at "rest". Eventually the expanding earth will intersect it.

In Figure 10b we see the same image, but with the arrows pointing inward indicating the earth stays a constant size and the test mass accelerates towards the earth until they intersect. In reversing the sense of the relationship between "matter" , in this case the earth, and the surrounding space, I bring the GEQ concept into agreement with what we observe and get rid of infinities. To complete the model I add a second test mass in 10b where this one has an added tangential velocity [$V = (GM/R)^{1/2}$] so it goes into orbit.

However, it is very important to keep in mind that while this is a convenient way of thinking about GEQ, the role of "space flowing inward" is an artifice. Space is actually "passive". All energy causing the action resides in matter.

This is different from the GR view in which both matter and space are active partners as previously described.

With this in mind, note that Figure 10b can also be represented by Figure 9 *so long as one keeps in mind* what causes objects to fall towards the central mass.

Let me comment on this again because it is so important:

A central tenet of GR is that space is an active "thing" in the same sense that matter is a "thing" with various physical characteristics. The relationship between "space" and "matter" is often described as, *"matter tells space how to curve, space tells matter how to move."*

In contrast a central tenet of GEQ is that space is passive, not a "thing" and matter does all the work. So while I can represent GEQ with the same "rubber sheet" model used in GR, they have distinctly different interpretations regarding "who does what to whom" and where energy-for-change lies. In GR it is shared between "matter" and "space"; in GEQ it resides only in "matter".

This says a lot about the similarities and differences among GR, GEQ and Newtonian gravity:

Neither GR nor GEQ talk about "gravitational forces" as used in Newtonian Physics. Instead:

a. GR says objects moving in space follow their "natural trajectories" because mass causes space to curve.

b. GEQ says objects moving in space "appear" to follow their natural trajectories *but do so with neither forces nor spatial curvature required!*

Chapter 6 deals with the underlined statement just above.

SUMMARY OF CHAPTER 6:

I don't have much to say about this chapter. You will just have to read it. As my friend Victor used to say, "it's Howdy Doody Time." Or "this is where the rubber hits the road." I have provided you with the all the rationale you need to understand in GEQ terms how orbiting objects orbit without central forces. But when presented with a concrete example, most readers have great difficulty in believing it because of our *certainty* of how we understand what we know about *direction!* If you are out in empty space and throw a ball, you understand it's time to say "Bye, bye ball," because that's the last you will see of it. But here on earth? Read on.

CHAPTER 6

Orbiting With No Force

The notion that orbiting bodies require neither forces nor spatial curvature to travel as they do challenges us. Our paradigm excludes such a possibility. Let me start with that weirdness.

Consider an observer standing on the earth with a test mass directly above him, "at rest" as in Figure 11.

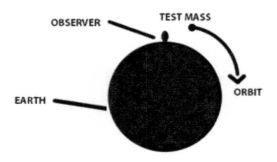

OBSERVER ON EARTH

FIGURE 11

I give it a push, so it has a constant velocity tangential to the earth's circumference. In the Newtonian view there is a central force acting on the test mass, causing the small mass to follow a curved path and go into orbit around the earth. GR replaces the central force concept by claiming space is curved in the vicinity of the earth. The small mass follows its "natural path" (Einstein calls it a geodesic) and likewise goes into orbit.

In both of these descriptions an external agency is involved in the orbiting of the test mass. In the Newtonian view there is a central force. In the GR view spatial curvature results from the effect of mass. The GEQ explanation is different. It claims the path of the test mass is due to the observer accelerating (being in a state of motion). Specifically, because he is accelerated, he sees the path as curved. No external agency is required for this result.

While this assertion is in accordance with the argument presented in chapter 3 and illustrated by Figures 4 through 10, it violates our understanding of reality, how the universe behaves. Logic is one thing; belief and comfort with belief are quite another.

All of this gets worse when I argue that logically, if there are many observers on the earth, distributed so that as the test mass passes from one observer's view on its curved path to the next, it can be deduced that the test mass will eventually come back into the view of the first observer. It can be described as "in orbit" around the earth. This is pictured in Figure 12.

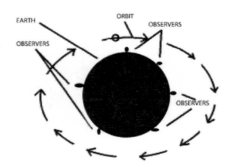

ORBITING TEST MASS

FIGURE 12

I can use the same argument from GEQ and note that if two objects are sent into identical orbits around the earth, but in opposite directions, they will eventually collide. One cannot claim this is just an illusion and not really happening. It has to be accepted as, at the very least, a possible explanation of how things work.

I can go even further and point out that this is consistent with the notion that an "at rest" observer near the earth will also see the small test object in orbit. His view is in a *shrinking coordinate system* (as indicated in Figure 13).

Although that observer is not accelerating in the usual sense, he still sees the mass in orbit.

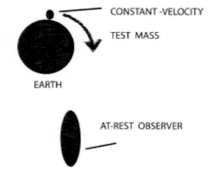

CONSTANT -VELOCITY

TEST MASS

EARTH

AT-REST OBSERVER

AT-REST OBSERVER, TEST OBJECT IN ORBIT, SHRINKING FRAME

FIGURE 13

There is yet another aspect of orbiting objects that brings into question what GR means by curved space. If space (or space-time) is curved, should not bodies in space maintain orientation analogous to how such objects orient in "flat space?" I am thinking specifically of angular orientation, where an elongated object (see Figure 14) maintains its orientation while moving.

ELONGATED OBJECT MOVING IN A STRAIGHT PATH

FIGURE 14

What is indicated is that any object with no angular momentum will maintain its orientation relative to, say, stationary distant stars.

In contrast, consider what happens in Figure 15. The elongated object possesses no angular momentum and maintains its orientation as if it were traveling in a straight line, not a curved path.

Notice that the orientation of the elongated object *always faces the same way*. This is an important difference compared to what is depicted in Figure 16,

where the object has had *angular momentum (rotation)* added to it. The net result in this second case is that the added angular momentum allows the object to travel with the same side always facing the central body.

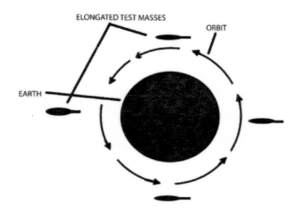

ELONGATED OBJECT IN ORBIT

FIGURE 15

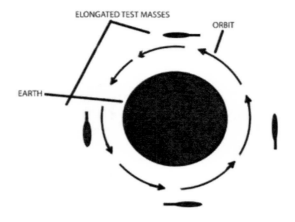

ROTATING ELONGATED OBJECT IN ORBIT

FIGURE 16

The point of course is that curved space is not quite analogous to flat space. I previously hinted at this difference in Figure 6, where I showed a small airplane that appeared to be moving at an angle relative to its axis.

Does this say that the concept of spatial curvature is somehow wrong? I think not, but it does suggest that as valuable as the idea is, it perhaps is not "real," but more of an apparent effect caused by our observational circumstances.

This last example, which explains tidal forces, uses the Newtonian view, but since the GEQ approach is identical to Newton's if taken to be *an inward flow of space* (see Figure 10b in chapter 5), the example suffices to show that GEQ can deal with this type of mechanical behavior. At the same time it makes clear that using GEQ *as an expansion of mass phenomenon* can lead to a more difficult analysis and is not always the best way to model GEQ. In later chapters, the effects of the expansion view, particularly using equation 6 (see equation derivations in chapter 2) will be examined, and one will for the first time be able to see the power of this new approach.

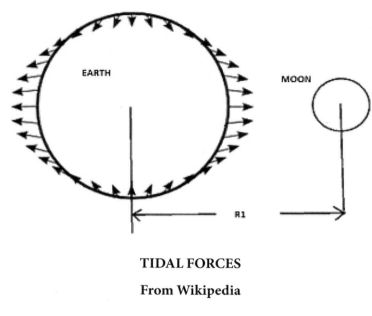

TIDAL FORCES

From Wikipedia

FIGURE 17

Figure 17 shows a diagram of forces from the moon acting on the earth viewed along its axis of rotation. While other forces (such as centripetal forces generated by the earth's rotation and effects from the Sun) are not explicitly shown, they do not qualitatively change the following explanation.

Note that the forces shown by the small arrow vectors are *towards* the moon on the side of the earth closest to that body and *away* on the opposite side. As the moon orbits around the earth the direction of the arrows will rotate with it.

Since the moon is in orbit around the earth, the forces at the distance R1 (the distance between the center of gravity of each body) cancel (outward force from angular momentum of the moon is equal and opposite to gravitational force at that distance). But the force on the nearer side is too intense causing the oceans to move towards the moon. The forces on the opposite are too low causing the oceans to bulge away from the center of the earth. Oceans at 90o from the moon experience inward forces and move inward.

What this shows is why we experience two high tides and two low tides each day as the moon rotates around the earth in one (roughly) 24-hour period.

While, as stated, this suffices to show that the expansion idea does satisfy the immediate requirement, it still is worthwhile looking at the solution in terms of an expansion model. The reasoning in that case, in which I assume the moon is in circular orbit and I treat the moon as a point mass (i.e., it doesn't expand), goes like this:

1. All energy resides in mass so the earth expands in accordance with the law, $acceleration = +GM/R^2$.
2. There are no gravitational forces.
3. All dimensions increase in size in such a way that the ratio of all sizes is time invariant from any observer's point of view.
4. Accordingly, an observer on the earth sees the distance R1 as time invariant, as in Figure 17.
5. However if the moon were momentarily "at rest" (not in orbit), an observer on the earth would see the distance decreasing between the surface of the earth and the moon.
6. To meet the requirements of items 4 and 5, it must be that the earth is bulging towards the moon in its nearest distance to the moon, and also bulging away from the moon at the furthest distance.
7. QED.

While this is conceptually correct, to verify it one would be required to go through a mathematical analysis of the model before it should be accepted. In view of this, it is clear that analysis of the model of Figure 10b is preferable. There is a lesson here: most of the time Newtonian methods are superior for analyzing phenomenon similar to this one, not involving long distances or velocities close. The same comment applies to the use of GR.

CHAPTER 7

Relation Between Geq And Gr

GEQ and GR are distinct theories, with very different roots, of what makes gravity work. For GR, the essential notion is that mass causes adjacent space to curve, that both space and mass are active elements in determining how gravity behaves, and that its actions are entirely deterministic to our limits of making accurate measurements. GEQ is very different in that it assumes that at bottom gravity's behavior is statistical in nature, directly caused by processes of subatomic particles and only appears deterministic at our levels of experience.

Further, all of the energy causing gravitational interactions is contained in matter. Space, so far as gravity is concerned, is a passive stage.

Both views take as guiding principles the teachings of Special Relativity. No material object can reach the velocity of light. One must consider time as a variable in describing behavior in a sense similar to the manner in which one uses spatial parameters.

GR has a very distinguished history. Over more than 100 years of extensive measurements of its predictions, none have thus far suggested it is incorrect. On the other hand, some phenomena have evaded explanation without introducing additional factors that are ad hoc in the sense they are arbitrarily assumed to explain some observed process with no theoretical basis other than that they work. While appearance of such "fixes" is the hallmark of a crises (need for new theory), that level has not been reached for GR. At most, those situations might suggest a need for some slight extension of the present boundaries of GR. However, the inability to include GR (gravity) as a member of the standard model of all forces might be considered a more important deficiency.

The situation for GEQ could not possibly be more different. It has no history beyond what appears in this book. There are no great crises in physics that demand its introduction (such was true for Special and General Relativity) and the motivation for even considering it is based on somewhat vague ideas about how the second law of thermodynamics interacts with the understanding of gravity's behavior. What is worse is the clear disagreement with GR about the role of space in the dance gravity presents to us.

If the only differences were the use of language and models in addressing solutions to mechanical problems, those differences might be excused. Either approach is equivalent to the other and in some cases one view is easier to work with than the other. But the differences go well beyond that when one gets to the topic of cosmology and how data is interpreted.

In cosmological models, as understood by the community, the role of space is taken to be an active element up to and including an expansion process, a process that is not bound by the speed of light and the plotting of relative velocities between distant galaxies that can reach or exceed that velocity. The underlying theory is called *cosmic inflation* and was first proposed by Alan Guth and other researchers in the late 1970s. It has since become the standard understanding of cosmology.

Since GEQ claims all energy for gravitational phenomena is resident in matter, it is difficult or impossible for GEQ and GR to both be correct. Any long- range data about velocities and distances, mainly from observations of Type Ia supernovae, are interpreted in the context of inflation theory and give rise to an understanding of the way the cosmos behaves. If the same data were viewed from the viewpoint of GEQ, the understandings no doubt would be very different.

One possible solution is to see GEQ as an alternate way of looking at gravity's behavior, but at a much shallower level. In that case, perhaps GEQ should have made its first appearance sometime around 1900 and Einstein could have used it as a guiding principle for his development of the Special and General Theories. If that had happened, he might have included statistical factors at the roots of GR and the inclusion of GEQ into the standard model would be clear.

All this is speculative. What is not speculative is that GEQ can address and solve many of the same problems as GR, including its very first triumph, the advance of the perihelion of Mercury. In addition, GEQ provides a solution for the bending of light and the behavior of spiral galaxies without the ad hoc inclusion of dark matter or MOND (modified Newtonian gravity). Perhaps most interestingly, it provides a model for investigating the suspicion that the earth has expanded over time. The analysis says indeed it has. At the same time, ,several experiments are described that would evaluate the GEQ notion of expansion, either supporting or falsifying it.

There is another way of looking at the relationship between GEQ and GR that is more radical than one would like because of the cosmological conflict.

That is, to take GR as a "black-box" description of gravity and GEQ as a description of the hidden inner operations or, so to speak, the "machinery."

The black box approach to understanding complex systems is, generally speaking, a very powerful analytic method for understanding and predicting causal behavior. But it does tend to suffer from assumptions made about how the system should be described when considering new phenomena that are on the edge of what is well-known and verified by experiment, that all important decider of validity.

If their relationship can be described this way, both views are valuable in their respective domains and GEQ has an important role to play in two different ways. First, it places space in a more properly limited role in its action (or lack of action) in how gravity behaves. Second, it brings into better focus the behaviors of a number of phenomena that from the GR point of view require the introduction of several new parameters (dark matter and dark energy). GEQ claims these are not required.

Having said this, let me make clear that I am not saying I think GR is wrong. I believe it remains a very powerful tool for uncovering the behavior of the universe. However, I think its application to cosmology, where space is described as a "thing" with energy and equivalence to matter in its actions, is wrong. I also think taking this point of view does not deprive the physics community of any useful understandings, including Alan Guth's idea of inflation, but instead places those ideas in a better perspective as models of behavior rather than literal truths.

In the sciences there is a principle called Occam's razor. According to that idea, if there are multiple explanations for some phenomenon, choose the simplest. Given our present state of knowledge, as an example, what is the simplest explanation for orbiting? Well, depending on your level of education, you might choose Newton's central force explanation, or if you have been inculcated into the stratified world of GR, you probably would claim spacetime curvature, but it is highly unlikely that you would choose what GEQ has to say!

If I want to change your choice and bring you into the world of GEQ, I have to show you a lot more evidence than I have so far. I must also provide something far beyond just more evidence; I have to present a necessity for this alternate choice because of some failures in the other explanations or some specific advantage to the GEQ approach under some circumstances.

The former of these is exactly what happened to bring Special Relativity (SR) (1905) and then GR (1915) into acceptance. In the context of Newton's views about gravity and the community's belief of how light propagated, certain difficulties appeared when measurement of light's motion through empty space did not support the belief of what was called the "luminescent aether."

On the whole I do not think I can demonstrate any failure of GR, although I do like the view that spatial curvature is only apparent and that space is passive in the dance. But I do not see these as failures so much as useful artifices to enable proper analysis of many difficult physical processes using the black box approach.

What I think I can do is show that in revealing the approximate nature of equation 7 and focusing attention on the small difference between equations 6 and 7, I can demonstrate solutions to a number of physical processes that GR, using its black box approach, cannot solve without introducing ad hoc phenomena strictly required for those solutions, where there is no a priori evidence that those phenomena have even a theoretical basis for their necessity. In other words, they are specifically assumed in order to address the difficulties required in the context of the existing theory.

This, of course, is exactly commensurate with the kinds of crises that appear in the sciences now and then, frequently preceding new ideas that end up advancing our knowledge of how the universe behaves and clarifying the inner machinery of the black box. That is what might be happening here. Does GEQ really describe the machinery of gravity? Maybe.

THIS ENDS PART 1

CHAPTER 8

Introduction To Part 2

The focus in Part 1was to clarify the behavior of local phenomena from the point of view of expansion and to demonstrate that those behaviors match those examined using Newtonian physics. I can well imagine the criticism that, with the possible exception of the thought experiment described in chapter 3, where the GEQ explanation of the goings on are somewhat more satisfying than those offered by the Newtonian view, I have done little more than offer an alternate explanation for gravity without offering any real motivation for accepting this new view. What advantage does it offer? What would compel any knowledgeable individual to choose this novel explanation over Newtonian methods?

In addition, except for a few comments about space-time curvature, I have hardly mentioned GR at all. Yet, I have the temerity to suggest that GEQ and GR both describe the same phenomenon from two different points of view! What is my justification for such a claim?

My justification derives from what appears in the next group of chapters, making it clear that the primary issue is *not* based on the similarities between GEQ and Newtonian gravity, but rather the *similarities between* GEQ and GR.

In the preceding chapters we have seen that the roots of the expansion hypothesis come directly from Newton's gravity law. This is a bit strange since we know from our present perspective that while that law is not wrong (after all I stated that it is one of the Truths in physics) and that we must accord it a very special position in physics, it is only valid under limited conditions.

To start the development of the expansion hypothesis using a law limited in applicability and to see it morph into a new law that may have broader applicability, simply by changing a negative sign into a positive sign, seems a bit unexpected.

On the other hand, since the purpose of Part 1 was to inform the reader about the properties of expansion, comparing its methods to those of Newtonian gravitational theory is justified by the need to show, as simply as possible, that this unfamiliar way of viewing things leads to the same conclusions as the older method. If this were not true there could be no claim of validity for this different way of analysis.

The development as presented does consider the properties of space and time in a limited way, sufficient enough to carry out the analysis presented in chapter 5 (where considerations of the role of coordinate systems affect our understanding of reality). Newton believed that "time was the same everywhere and that the transmission latency of any gravity field was zero. The effects were instantaneous regardless of any distance separating the source (some sizable hunk of matter) and the recipient (say a test mass). Einstein showed in his development of Special Relativity that these assumptions were untenable. Therefore, one must view the developments in previous chapters as incomplete.

What I can add to the discussion relates to the ability for the expansion hypothesis to make a statement that General Relativity cannot. That is, *gravity and acceleration are identical.* In GR, Einstein makes two statements related to this assertion, but neither of them goes as far as the above. You will read shortly about what he does say, but for now note that his idea, called **principle of equivalence**, stops short of that claim even though it is sometimes described as doing so. What he did claim was that for **weak equivalence**, inertial and gravitational mass have the same magnitudes *at least in*

72

the circumstance of any uniform gravity field, and for **strong equivalence** all laws of physics are the same including in gravity fields of any form. Newton knew the first of these was true but didn't know why; he never commented on the latter.

Now, in the following chapters, comes the exploitation of that small difference in metric definition that, while small in magnitude, makes possible very large claims concerning the relationship between matter and space, the primary target of Einstein's geometric theory of gravity, GR. Since the targets are the same and the results are at least nominally the same, I see no reason not to claim that they describe the same phenomena with different approaches.

Briefly, here are the topics for the next 7 chapters.

Chapters 9 and 10 show that GEQ methods can be used to explain the perihelion advance of Mercury (chapter 9) and the bending of starlight (chapter 10) through considerations of the small difference in metrics implied by equations 6 and 7. The importance of this is that both GR and GEQ can solve these anomalies without further ad hoc assumptions.

For Einstein, the solution of Mercury's advance was particularly important. He demonstrated it *before* he had fully developed his ideas and it showed he was on the right track. If the solution to the advance of Mercury's perihelion was Einstein's most amazing experience, mine was the expansion of the earth but it did not give me "palpitations," but rather it allowed for a very good night's sleep. *Chapters 11 through 15* cover five topics: expansion of the earth (chapter 11); NASA's anomalous acceleration of Pioneers 10 and 11 (chapter 12); cosmological models (chapter 13; spiral galaxy anomaly chapter 14); and the relation between the Hubble constant and GEQ (chapter 15).

In considering the first of these, expansion of the earth, the major attempts at solution to date use unframed measurements with no guiding principle and are unable to evaluate measurement results. The GEQ approach provides a theoretical framing. It eliminates confounding variables and shows that earlier flawed methods managed to come close to a proper answer but with no way to evaluate their conclusions. Of particular importance, the experiments described would provide definitive evidence of the validity (or not) of the GEQ hypothesis and the mechanism underlying the earth's increase in size.

The second topic (anomalous acceleration of the Pioneer probes) analyzes the three anomalous acceleration terms in the Pioneer series of deep space studies. It shows that all three terms can be accounted for by the small difference in metrics between equations 6 and 7.

The third topic (cosmological models) outlines the GEQ approach to cosmology.

In the fourth of these (the spiral galaxy problem), the GEQ methodology is used to show that neither *dark matter* nor modified Newton's gravity law, *MOND*, are needed.

The last of these, Chapter 15, is perhaps the most controversial topic of the book. It contradicts the interpretations of cosmology. Specifically, that spatial expansion exceeds the velocity of light. The assumption in GEQ, that all expansion energy is resident in matter, results in a very different interpretation of type Ia supernova data. I will say no more here about this important issue for it requires considerable explanation and analysis.

SUMMARY OF CHAPTER 9

If you read the first six pages of this chapter you will get the history of both the advance of the perihelion of Mercury and why the explanation of the phenomenon was so important to Einstein. In brief, his development of GR implied that the distance around a circular orbit was slightly greater than 2pi*R, which is the usual (Euclidean) way of thinking of the relationship between any circle and its radius, R. According to his new understanding, Euclidean geometry *did not* apply to the way space worked; one had to use Riemannian geometry. The consequence was that Mercury had to travel a bit further on each rotation around the Sun so that was why the advance. His solution was entirely mathematical and very complicated, so in the text, I only outline it

The GEQ method is much simpler than the GR approach, but still mathematical. It recognizes that the usual definition of length (length = distance times light speed) is only approximately true. This definition corresponds to equation (7), while the true definition should be equation (6). The GEQ solution shows that the difference between equation (6) and (7) is the same as the difference for Mercury's path length as it is using Euclidean geometry instead of Reimann geometry. In either case, the solution has to consider that the orbit is elliptical, not circular, so there are a number of complicating factors, but in the end, it is shown that the GEQ solution is far simpler than the GR approach.

KEY CONCEPTS FOR CHAPTER 9

1. Time Dilation from Special Relativity

$$\Delta t_B = \frac{\Delta t_A}{\left(1 - \frac{v^2}{c^2}\right)^{1/2}} \qquad\qquad (8)$$

Observer in "B" coordinate system with velocity v relative to A sees time intervals in "A" system as longer than in his own system.

1. For Schwarzschild metric *not* in orbit:

$$t_B = t_A \left(1 - \frac{(2GM)}{Rc^2}\right)^{1/2}$$

That is read as, "observer B, at rest above   earth, sees observer A's time intervals on the earth go slower than his.

2. For circular orbit in Schwarzschild metric:

$$t_B = t_A \left(1 - \frac{(3GM)}{Rc^2}\right)^{1/2}$$

That is read as, "observer B in circular orbit around the earth sees observer A's time intervals on the earth go slower than his

4. Mass/energy relation: Rest Mass: $E = Mc^2$

   Not at rest mass $E = Mc^2/(1 - v^2/c^2)^{1/2}$

That is read as, observer A in circular orbit around a central mass sees observer B's time in distant space move faster than in his system. Or time intervals for him are slower than B. This _does_ include time dilation from SR.

CHAPTER 9

Advance Of The Perihelion Of Mercury

"This discovery was, I believe, by far the strongest emotional experience in Einstein's scientific life, perhaps in all his life. Nature had spoken to him. He had to be right. "For a few days, I was beside myself with joyous excitement." Later, he told Fokker that his discovery had given him palpitations of the heart. What he told de Haas is even more profoundly significant: when he saw that his calculations agreed with the unexplained astronomical observations, he had the feeling that something actually snapped in him. —Abraham Pais ("Subtle is the Lord...": The Science and Life of Albert Einstein 1982)

The planets of the Solar System have played a central role in Western gravitational theory from the time of the Greeks up to the present. Now, gravitational theory has morphed into asking larger-scale questions of how the universe behaves.

In 1630 Johannes Kepler published his three laws of planetary motion. Newton followed by publishing his gravitation laws in 1667. These teachings became the theoretical underpinnings for all astronomical research until Einstein appeared on the scene almost 300 years later.

One of the early major astronomical triumphs was Edmund Halley's prediction in 1705 that a comet observed in 1531, 1607, and 1682 would return to view in about 75 years (1758-59). A French astronomer, Alexis-Claude Clairaut, used Newton's law to predict a more precise date and arrived at a solution of mid-April 1759, only 33 days off from when it did occur.

What is of great importance is that Clairaut's prediction included perturbations in the orbit caused by Jupiter and Saturn. This showed that orbits were modified by other planets in the Solar System, an observation that turned out to play a large role in explaining Mercury's behavior and thus enabled predictions about other bodies not known of at the time.

As if to drive this point home, another astronomer, William Hanover, discovered a seventh planet in 1781, Uranus, that followed a trajectory *almost* predicted by Newton's law, but slightly off.

In 1841, Urbain Le Verrier, a brilliant polymath, who ended up in astronomy by accident (he had planned to be a chemist) began a detailed study of Mercury's orbit attempting to bring Newtonian predictions into agreement with observations; this was considered to be a critical test for Newton's theory. By 1845 he had calculated Mercury's perihelion to within 16 seconds.

However, he was unhappy with this error and never published his results.

He turned his attention to the deviation of Uranus's orbit from the Newtonian prediction and focused on the idea that there was yet another planet outside the orbit of Uranus. In 1846 he announced the presence of an eighth planet, Neptune, which was sighted by the Berlin observatory within 55 arc minutes of where Newton's law said it should be. The director of the observatory exclaimed that this was *"the most outstanding conceivable proof of the validity of universal gravitation."*

Le Verrier, with this success under his belt, was appointed director of the Paris Observatory (1854) and returned his attention to Mercury in 1859. He reasoned that if an unknown planet had caused the deviation in Neptune's orbit it was likely the same cause for Mercury's behavioral discrepancy from what Newtonian calculations predicted. Accordingly, he turned his attention to that new "*missing*" planet.

As Director he had at his disposal a large staff to carry out detailed calculations and soon had in hand an accurate estimate of what Newton's law said Mercury's orbit should be. He focused on the timing of the perihelion's precession, measured to be 565 seconds of arc per Earth century (observation of 14 transitions from 1697-1848) while his calculations said it should be 527 seconds. He was off by 38 seconds of arc.

He calculated the mass and orbit of a planet closer to the Sun than Mercury and quickly realized it had to be a group of smaller bodies, not a single one, which would have been easily observed during an eclipse. However, an amateur French astronomer convinced him he had observed such a planet. Le Verrier reported this result as fact, the popular press named the new planet Vulcan, and once again Le Verrier and Newton's law were acclaimed by the public.

Unfortunately for him, in spite of many attempts to sight Vulcan, neither he nor any other astronomer found it. Le Verrier died in 1877 still believing it existed and that Newton's gravity law was correct.

In 1895 Simon Newcomb, an American Canadian astronomer, refined the measurements of Mercury's orbit and demonstrated that the extra advance of its perihelion was 43 seconds of arc, a bit more than 38 seconds, and that is essentially identical to the current value.

From 1895 through 1905 interest in perihelion measurements switched to questioning if perhaps Newton's gravity law was not quite correct. While a number of ad hoc suggestions were made, none were taken seriously. Since this was not the only troublesome finding during this time period, it began to look as though a crisis in physics was imminent. The community's attention turned to other more important matters until after 1905.

A model of how Mercury's perihelion behaves is diagramed in Figure 18 where the angular magnitude is greatly exaggerated for clarity. When and if Einstein or anyone else were to supply an explanation for the extra 43 seconds of arc, this was the data and model they had to work with.

Advance of the Perihelion

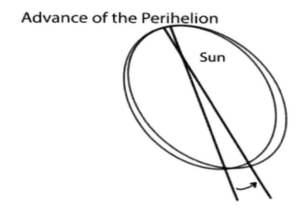

Sun

ADVANCE OF THE PERIHELION OF MERCURY

FIGURE 18

In 1905 Einstein published four groundbreaking papers: The first was on the photoelectric effect; the second on Brownian motion; the third on special relativity; the fourth on mass-energy equivalence. The year 1905 is often referred to as Einstein's Annus Mirabalis. It is the third of these, the one on special relativity, that provides us with the thread to the motion of Mercury.

The formal name of that paper was *On the Electrodynamics of Moving Bodies*. The paper clarified how light propagates through empty space, that the laws of physics are the same for all observers not accelerated (in inertial states), and that all observers see the velocity of light the same way and as constant in magnitude.

What it did not do was deal with gravity in any manner; the universe this paper described was devoid of any masses! From the moment he wrote it Einstein knew this notion of his, successful as it was (greeted with acclaim), required still more; another complete development that did include matter, a new theory of gravity. While this new theory still had to agree with Newton's gravity ideas where they had been shown to be correct, it had to go beyond those ideas and mesh together with the new concepts of special relativity.

His work had just begun. Indeed it took over ten years before he accomplished his goal between 1907 and 1915. It was a difficult trip and caused him to retrace paths he thought fruitless. Finally, he reached his goal, publishing an enormous paper over 75 pages long. He called it *The Foundation of The General Theory of Relativity.*

The quote at the beginning of this chapter refers to a bright spot in his frustrating path that occurred in 1914, shortly before he completed and published his new theory. What his solution gave him, and what he said in a lecture given in 1915 was, *"The calculation yields, for the planet Mercury, an advance of the perihelion of 43" per century, while the astronomers indicated 45sec. of arc± 5 the unexplained remainder between observations and the Newtonian theory. This means full compatibility."*

The actual fact is that his method of calculation was not entirely clear and led to a "friendly war" between Einstein and Karl Schwarzschild (who is credited with being the first to solve any of Einstein's complex field equations).

Schwarzschild, shortly after Einstein's solution, put forth his own method based on Einstein's work. He called it "an exact solution." He also wrote Einstein a letter in which he commented, *"...such a conclusive clarification of the Mercury anomaly. As you see, it means that the friendly war with me, in which in spite of your considerable protective fire throughout the terrestrial distance, allows this stroll in your fantasy land." (Dec 1915)*

Disregarding this friendly battle over methods, the essence of Einstein's insight was that the Newtonian solution to orbits did not consider a small deviation in actual path length, thus causing the perihelion to not quite repeat its time occurrence on each cycle. The equation he derived took the form:

$$\Delta\sigma = 24\pi^3 \frac{a^2}{T^2 c^2 (1-e^2)}$$ (9)

Where a = semimajor axis or distance from the Sun; T = period around the Sun, c= speed of light and e = eccentricity of Mercury's orbit.

The (1-e2) term in the denominator is a correction for the difference in path- length for an elliptical orbit, the actual path Mercury follows, and a circular path, which is always the assumption for calculations.

For the perihelion of Mercury, a = 5.8 x 1012 cm; e = 0.206, M of Sun = 1.989 x 1030 kg, and T = 7.6 x 106 seconds (approximately).

I will not attempt to reproduce any of the GR solutions here because they are all lengthy and complex and generally run to over forty mathematical steps. A good example is given by Anatoli Vankov, available on-line, titled, *Einstein's Paper: Explanation of the Motion of Mercury from General Relativity Theory.*

If one uses his equation to calculate the perihelion advance for each revolution of Mercury around the Sun one obtains approximately:

$\Delta\sigma$ = 4.8 x 10^{-7} radians/revolution for the circular orbit   or

$\Delta\sigma$ =5 x 10^{-7} radians/revolution for the elliptical orbit.

The usual conversion of radians/revolution for that time period to seconds of arc per century is **5 x 10-7 x (180/pi) x 3600 x 415 = 42.8 arc seconds.**

GEQ SOLUTION

Just as Einstein was excited to show his methods finally solved this mystery, so am I to show that my approach explains this discrepancy from the Newtonian solution. This is how GEQ approaches the problem.

For convenience let me repeat those all-important equations:

For convenience let me repeat those all-important equations:

$$t_{eq} = \frac{R}{c}(1 - \frac{R_{miin}}{R})^{1/2} + \left(\frac{R_{miin}}{c}\right) \ln\left[\left(\frac{R}{R_{min}}\right)^{\frac{1}{2}} + \left(\frac{R-R_{min}}{R_{min}}\right)^{\frac{1}{2}}\right] \tag{6}$$

$$t = \frac{R}{c} \text{ or } R=c*t \tag{7}$$

(Equation (6) represents the exact size of the sphere at any time (R = c*t$_{eq}$); and

Equation (7) tells me the approximate size of the sphere at any time. (R = c*t) and note, the mass term "M" is missing in equation (7)!

If you recall, I pointed out that the derivation that led to equations 6 and 7 did not explicitly include time dilation from Special Relativity (SR). It also does not include the mass/energy relation for observers in relative motion. So, to the extent those factors play a role in the advance of the perihelion, one might suspect that any attempt to derive the magnitude of the advance would fall short of supplying a proper answer. Indeed, that suspicion is correct. Nevertheless, it is worthwhile to go through that analysis to demonstrate the method. After this initial exercise, I will repeat the derivation in greater detail including time dilation effects and the mass/energy correction.

To begin, take notice of Kepler's third law for the orbit of any planet: *The square of the orbital period of a planet is directly proportional to the cube of the semimajor axis.*

In current mathematical form this is written as:

$$Tm = 2\pi\left(\frac{a^3}{GM}\right)^{1/2} \qquad\qquad (10)$$

Where: Tm = period of path; a = semimajor axis.

In the case of a circular orbit this is the same as:

$$Tm = \frac{2\pi a}{V} \qquad\qquad (11)$$

Where V is the orbital velocity

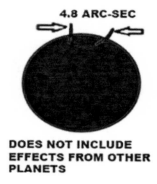

4.8 ARC-SEC

**DOES NOT INCLUDE
EFFECTS FROM OTHER
PLANETS**

ORBITAL ADVANCE OF PERIHELION FOR CIRCULAR ORBIT

FIGURE 19

Figure 19 shows the circular orbit with the period (Tm) marked out and the actual over-run interval that represents the perihelion advance as measured.

According to equation 6 the advance interval arises from the Newtonian assumption that the radial distance, a, is properly defined by $a = c^*t$, while GEQ using equation 6 says it is actually given by $a = c^*(t_{eq})$.It turns out that the quantity teq/t is always greater than unity, so the path length (period>t) for each orbit has to travel a bit extra than 2pi for each rotation.

That is: $Tm = \frac{2\pi c t_{eq}}{v}$ Mathematically, expressed as radians, this is given by:

$$\frac{\left(Tm\frac{t_{eq}}{t}-Tm\right)}{Tm}$$ radians.

In the case for Mercury this evaluates to 4.4x 10-7 radians for the circular orbit; and 4.6 x 10-7 radians for the elliptical orbit. Converting to per century, one has about 39 arcseconds, too low as expected.

What is missing are corrections for time dilation for a body in circular orbit and time dilation from SR. The calculation is as follows for Mercury (where the Schwarzschild metric is used since it supplies the correct time dilation term).

The Schwarzschild metric for this application takes the form.

$$t_B = \frac{t_A}{(1-\frac{sGM}{Rc^2})^{1/2}}$$

Step 1: Calculate t_{eq} = **193.16383843 sec** (note t = 193.16375381 sec)

Step 2: Calculate t_{eq1} using the Schwarzschild metric,

$$t_{eq1} = \frac{t_{eq}}{(1-\frac{sGM}{Rc^2})^{1/2}} = \textbf{193.1638458 sec}$$

Step 3: Calculate advance for circular orbit,

$$\frac{\left(Tm\frac{t_{eq1}}{t}-Tm\right)}{Tm} = 4.76x10^{-7} radians$$

Step 4: Calculate the extra radians per revolution for the elliptical path,

$$\frac{\left(Tm\frac{t_{eq1}}{t}-Tm\right)}{Tm(1-e^2)}$$ radians = **4.97 x 10-7 radians** $\cong$ **5x10-7 radians**

Step 5: Calculate the advance for one year (415 revolutions),

5x10-7 x 180/pi) x 3600 x 415 = 42.8 arc seconds

Applying the same methodology to calculate the advance of the perihelia of the Earth and Venus, one obtains the following table comparing calculated values to measurement values.

TABLE I

COMPARISON OF PERIHERLION ADVANCES

PLANET	ADVANCE CALCULATED PER 100 YRS (GR)	ADVANCE CALCULATED PER 100 YRS (GEQ)	*ADVANCE MEASURED PER 100 YRS
MERCURY	42.8 arc-seconds	42.8 arc-seconds	43.1+/- -0.5
VENUS	8.6 arc-seconds	8.7 arc-seconds	8.0+/- 5.0
EARTH	3.9 arc-seconds	3.8 arc-seconds	5.0+/-1.0

Now finally, let me point out something interesting that suggests GEQ might be on the right track. If Kepler's third law is correct, and we have no reason to believe otherwise, then Nature should have adjusted the period so *the square of the* orbital period *of a planet is directly proportional to the cube of the semimajor axis.* If that adjustment had occurred there would be no perihelia advances!

The fact that those advances are present indicates that there is a monotonic increase in the radius that GEQ predicts, where the magnitude of that increase is governed by the difference between (teq-t). This follows because, according to GEQ, the metric of length is increasing in time. But the only portion of that increase effecting our measurements is that small increment.

CHAPTER 10

Bending Of Starlight By The Sun

As early as 1801, Johann Georg von Soldner published a paper noting that Newtonian gravity implied light was bent by the Sun. In 1911, long before he completed his work on the general theory (GR), Einstein likewise published a paper predicting the magnitude of the bending effect. He stated that it should be about 0.867 arc seconds as starlight passed by the Sun on its way to earth. Note in particular that this was about four years before he published his paper on General Relativity and about three years before his happy prediction about the advance of Mercury's perihelion. The timing of his prediction is of interest as subsequent attempts to measure the bending of light during solar eclipses failed. Due to this and other interruptions, the bending effect was not able to be properly measured until 1919.

In 1912, an expedition was formed by a Brazilian group to use an eclipse of the Sun to measure the bending of starlight. However, bad weather resulted in no useful measurements. It would have been disastrous for Einstein had the efforts been successful as his magnitude was off by a factor of two. In 1914, Einstein had completed enough of his work on GR to realize that his prediction was only half of the proper value. He published a new paper correcting his error. That same year a second expedition was formed. Organized by Erwin Finlay-Freundlich and William Wallace Campbell, the expedition travelled to the Crimea to observe the solar eclipse of 21 August. The start of World War I and bad weather combined to result in failed measurements again.

Enter Sir Arthur Eddington. Eddington's interest in general relativity began in 1916 (during World War I). He read papers by Einstein (presented in Berlin, Germany, in 1915) sent by the neutral Dutch physicist Willem de Sitter to the Royal Astronomical Society in Britain. Eddington, later said to be one of the few people at the time to understand the theory, realized its significance and lectured on relativity at a meeting of the British Association in 1916. He emphasized the importance of testing the theory by methods such as observing light deflection during eclipses. The Astronomer Royal, Frank Watson Dyson, began to make plan for the eclipse of May 1919, which would be particularly suitable for such a test. Eddington also produced a major report on general relativity for the Physical Society published as, *Report on the Relativity Theory of Gravitation* (1920).Eddington lectured on relativity at Cambridge University, where he had been professor of astronomy since 1913.

Wartime conscription in Britain was introduced in 1917. At the age of 34, Eddington was eligible to be drafted into the military, but the university obtained him an exemption on the grounds of national interest. This exemption was later appealed by the War Ministry. At a series of hearings in June and July 1918 Eddington, who was a Quaker, stated that he was a conscientious objector based on religious grounds. At the final hearing, the Astronomer Royal, Frank Watson Dyson, supported the exemption by proposing that Eddington undertake an expedition to observe the total eclipse in May of the following year (in order to test Einstein's General Theory of Relativity). The appeal board granted a twelve-month extension for Eddington to do so. Although this extension was rendered moot by the signing of the Armistice in November, the expedition went ahead as planned.

In actuality, two expeditions were formed: one to the Island of Principe off the coast of Western Africa and a second to the town of Sobral in Brazil.

Unfortunately, the photographs from Sobral were blurred and unable to supply useful data, but photographs from Principe were successful. This group was headed by Eddington and although the morning of May 29th was rainy, cloudy, and windy, the weather cleared by noon and Eddington wrote:

> *"The rain stopped about noon and about 1.30 ... we began to get a glimpse of the sun. We had to carry out our photographs in faith. I did not see the eclipse, being too busy changing plates, except for one glance to make sure that it had begun and another half-way through to see how much cloud there was. We took sixteen photographs. They are all good of the sun, showing a very remarkable prominence; but the cloud has interfered with the star images. The last few photographs show a few images which I hope will give us what we need ..."*

Later on June 3, Eddington developed the plates and wrote in his notebook the important words:

"...one plate I measured gave a result agreeing with Einstein."

The plate showed a value of 1.75 arc seconds, just as Einstein had predicted. This was the deflection of the star position from where nighttime sightings placed the star when the Sun was absent. Since the distant stars were considered stationary, the deviation in position was attributed to the effect of the Sun's mass! While some critics questioned the accuracy of the measurements, the experiment was repeated at the Lick Observatory during an eclipse in 1922. The results agreed with the earlier study by Eddington. Many similar measurements have been made since then, with the primary purpose of obtaining a more exact value.

When asked what he would have said if his prediction had been wrong, Einstein said: *"Then I would feel sorry for the dear Lord. The theory is correct anyway."*

Now, why did Einstein get it wrong in 1911? What made him correct the prediction in 1913? It is worthwhile taking a look at the sequence of his calculations, at least to a nominal extent. The situation is modeled in Figure 20.

I will outline the steps for Einstein's first and second calculations and provide the GEQ solution in full detail.

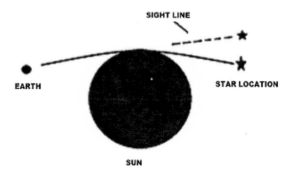

Bending of Starlight FIGURE 20

First, the Newtonian solution. In Einstein's earlier reasoning, which was wrong, one only had to change the velocity V to the velocity of light c to get the correct answer. Note in Figure 20 that the deflection in the path of the test mass is symmetric, the same deflection on approach to the Sun and from the Sun to the earth. Therefore, one can calculate the net deflection on just one side and double the result.

However given the direction of the forces in the path from the Sun to the earth, it is clear that the speed of the particle will slow down. But a photon by definition moves at the speed of light and *cannot* slow down (from SR, the velocity of light is a constant for all observers). Thus, the behavior of a particle must be different from a photon. One can argue that as the particle moves *towards* the Sun from the star it speeds up by the same rate. The effects cancel and I can ignore them. Let's do that.

Let us start by just considering the deflection as the mass m approaches the earth from the Sun.

When a force F acts on a mass m it results in the small particle accelerating along the direction of the force. It is simply:

$$\text{Acceleration} = -\,GM/R^2$$

Here I am using Newton's Law, unmodified.

Since acceleration is the time derivative of velocity

(i.e. dV/dt = acceleration), I can calculate the total change of velocity over the distance from closest approach to the center of the mass of the Sun to the earth by integrating the above equation over the time it takes the test mass to get from the Sun to the earth.

I can do this integration using the same method as in the derivation of equation (3), but here since I am assuming the force is attractive, I obtain:

$$Velocity = \frac{dr}{dt} = \left(\frac{2GM}{R}\right)^{1/2}$$

Here is the first difference between using a small test mass instead of a photon. Velocity is a vector quantity and can be resolved into two spatial components: one component parallel to the x axis and one component parallel to the y axis. We are only interested in the y component causing the test mass to move downward in the figure. That quantity is given by:

$$\tan\theta = \theta = \frac{GM}{RV^2}$$ Diagrammatically,

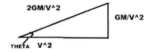

Let V go to c, double this value, and I get:

$$\theta = \frac{GM}{Rc^2} + \frac{GM}{Rc^2} = \frac{2GM}{Rc^2} = 4.5 \times 10^{-6} \text{ radians}$$

Multiplying this by (180/(pi)x3600 = 0.876 arc-seconds, which is half the proper value.

$$: \theta = \frac{2GM}{Rc^2} + \frac{2GM}{Rc^2} = 9 \times 10^{-6} \text{ radians} = 1.75 \text{ arc-sec}$$

This result is variously explained as the first term is due to the same factor as in the Newtonian solution and the second term is from space-time curvature **or** the first term is from spatial perturbation of the metric and the second term from time perturbation of the metric. Neither of these descriptions provides much help from an intuitive/physical view.

For that let me detail the GEQ explanation.

91

GEQ SOLUTION

Let me start with a model guiding the analysis. This is diagramed in Figure 21.

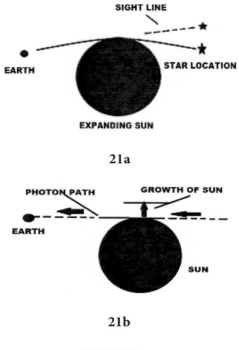

21a

21b

FIGURE 21

Consider the point of view for an observer located *on the Sun*. Aside from the issue of his turning to ash, it is equivalent to the observer on the earth indicated in Figure 11 (chapter 6) where the observer is radially accelerated. Because of his acceleration he sees the beam of light traverse a curved path both as it approaches and leaves as indicated in Figure 21a. Since the velocity of light is a constant, the deflection seen by the observer on the earth is the sum of two terms that are identical. The magnitude (deflection) of the terms must be determined from the growth of the earth during the elapsed time for a photon to travel from the Sun to the earth. Since the curvature is symmetric, the proper answer is simply double that value.

This can be found from equation (3) (chapter 2) for the velocity of expansion for an arbitrary sphere, written in the form.

$$Velocity = \frac{dr}{dt} = (c^2 - i\frac{2GM}{R})^{1/2} \tag{3}$$

Where the imaginary term represents the radial growth of the sphere for one-half of the total expansion as indicated in Figure 21b and the simple diagram just below.

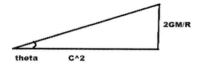

theta   **c^2**

$$2\tan\theta = 2\theta = \frac{2GM}{Rc^2} + \frac{2GM}{Rc^2} = 9 \times 10^{-6} \text{ radians} = 1.75 \text{ arc-sec}$$

Note that unlike either the Newtonian or the GR solutions, the entirety of the "bending" is assigned to the background framing (in this case the Sun), not to the photon that is not deflected in its path.

Now the process can be understood physically from Figures 21a and 21b.

Now the process can be understood physically from Figures 21a and 21b.

SUMMARY OF CHAPTER II

Over the years many people, including some scientists, have postulated that the earth has increased its size over time. A major clue suggesting this is the outline shapes of the continents that appear as though they were joined together into a supercontinent called Pangea. In 1912, a German scientist (Alfred Wegener) proposed that they had been connected in the past, and somewhat later fossils discovered on the western boundaries of Africa matched those on the eastern boundaries of the Americas indicated that it had formed about 350 million years ago and broke up about 175 million years ago. However, as more was learned about continental drift, his explanation was rejected as a solution, but the fossil data and the coastline contours remain as unanswered puzzles. Because of this, the notion that the earth is expanding in size has gained some traction, and efforts have been made to measure the effect.

The methods used to test the hypothesis have used various approaches focusing on changes over a number of years of geodetic sites with known elevations above sea level and where local geological instabilities are unlikely to have confounded data over the time intervals of interest. The results of these studies are that if a growth exists, it must be in the range of 0.35 +/- 0.47 mm/year.

Unfortunately, since they did not have any theory or model to guide them the final conclusion was, " ----- *that space-geodetic observation does not support the Earth expansion hypothesis."* In effect the studies do not meet the requirements of a scientific proof.

According to GEQ the expansion does exist and is entirely due to the difference between equations (6) and (7) that denies length is exactly determined by the time-lapse for a pulse of light to travel 1 meter. The deviation is very small but for a mass as large as the earth the magnitude is great enough to be measured under the right conditions.

To test this the chapter predicts the results of a series of measurements of height changes where the measurements are carried out for an extended period of time. These measurements are similar to the Pound Rebka series that were designed to test Einstein's prediction that photon energy increased when they "fell" in a gravity field. This is the gravitation time dilation effect discussed in chapter 4.

Those studies agreed with Einstein's prediction but it was clear that neither the data of those nor follow-on more sensitive studies would have shown the small "growth" phenomenon described in this present chapter.

The tests proposed in this chapter are actually easier than the Pound Rebka series but complicated by the need to carry them out for extended periods of time thus requiring long-term stability of the measurement devices. Given that the predicted outcomes are found to be correct, two things would be demonstrated: Evidence for the expansion (GEQ) hypothesis; and a reference standard for evaluating expansion results for the earth's size.

The remainder of the chapter assumes the expansion exists and predicts what the parameters would be if the distance between the tip of the nose on South Africa to the coast of South America at present bit more than 3000 miles, was reduced to 3 miles over a period of 175 million years. This would correspond to the apparent change based on fossil data.

What becomes apparent is that the definition of what is meant by "time interval" is a strong determinant of the final values because time intervals in the past were not the same as today, so the "actual" time lapse in years lies somewhere between 175 million years and 147 million years, depending on the definition used. The final conclusion is that the calibration data agrees strongly with the fossil data when adjusted for changes in the value of time intervals over 175 years or so.

Note from our point of view (assuming we were resident on the earth 175 years ago) there would be no change in dimensions because any changes would apply to our bodies and our measuring tools as well as to the earth.

CHAPTER II

The Expansion Of The Earth

Over the years many people, including some scientists, have postulated that the earth has increased its size over time. A major clue suggesting this is the outline shapes of the continents that appear as though they were joined together in the past. Figure 22 clearly indicates that the western boundary of Africa could nest into the eastern boundary of the Americas. The earlier form is a supercontinent called Pangea.

before   after

Stock Image (123RF.com)

FIGURE 22

Alfred Wegener, a German scientist first used that name in a 1912 paper in which he also set forth his theory of continental drift. He expanded on his ideas in a later paper (1915). While his idea was at first not accepted, discovery of fossils on the western boundaries of Africa that matched those found on the eastern boundary of the Americas eventually convinced the scientific community he was correct. According to fossil evidence, Pangea must have formed about 350 million years ago and started its break-up about 175 million years in the past.

The essential explanation according to the most widely held view in the recent past was that there is continental drift caused by subductions in tectonic plates deep in the earth and these subductions cause surface motions at continental scales causing Pangea to expand into its present geometry. There are definite difficulties with this explanation in that what we now know about plate tectonics does not support that view. Despite this, some scientists cling to the belief that what is going on deep in the earth is causing continental drift. One fact however is not disputed: Pangea did exist in the past and an adequate explanation for its evolution into its present form is yet to be determined.

Because of this, the notion that the earth is expanding in size has gained some traction and efforts have been made to measure the effect. One group doing so appeared in the publication, Geodesy and Geodynamics in 2015 by a group headed by Wenbin Shen et al. The paper is titled, *Evidence of the expanding Earth from space-geodetic data over solid land and sea level rise in recent two decades.*

Their results were typical of other researchers. According to their analysis, the expansion of earth's solid parts (if it exists) has been on the order of 0.35 mm/ year over the past several decades. The measurements are difficult and easily confounded by other factors such as melting of icecaps and other thermal expansion issues. They used two different methods for determining any earth expansion: Average changes in elevations of 629 geodetic sites as indicated by satellite observation telemetry, and consideration of oceanic expansion due to thermal expansion and melting of icecaps. Combining all their data gave them an approximate global expansion of about 0.35 +/- 0.47 mm/year.

Some of the details of this study are of great interest. Starting with 1572 geodetic data sites, they combined those sites close to one another ending up with 845 sites. They further removed all those showing large vertical velocities and the remaining 629 sites were used in the analysis stating, "*Absolute values of vertical velocities of some stations are relatively large. We consider that a too large vertical velocity should be related to local events rather than global expansion. Such stations are removed from our calculations. Stations locating in orogen belts are also removed since these vertical velocities of these stations are more likely related to local deformations but expansion.*"

They <u>do not</u> provide any further information about the altitudes of the removed or retained sites and as will become apparent, that factor is of prime importance. Their final conclusion was, "*…we note that using space geodetic data (GPS sites over the globe) Wu and his colleagues suggested that space-geodetic observation does not support the Earth expansion hypothesis*

While they are correct that their results fail to support the expansion hypothesis, if they had sorted their data by elevations their conclusions might have been very different, even if they didn't understand why elevation played such a large role in magnitudes.

GEQ SOLUTION

Now let me address this issue from the point of view of the expansion hypothesis (GEQ). According to the ideas put forth in GEQ, any time-varying measurements between sites at different elevations should show strong correlation with height difference because the factor allowing any difference at all to be observed is the *distortion parameter* (defined in the next several paragraphs) without which no time-varying differences in elevations would be observed regardless of heights above the reference level. Based on the results of direct calculations and the Pioneer 10/11 studies (see chapter 13) we have reasonable confidence that the earth has _distortion acceleration_ of about 1.4 x 10-5 cm/sec2 at sea level if measured relative to distances >>20 AU. For shorter height measurements the relative magnitude decreases roughly proportional to distance. Expressed as a constant expansion velocity for the earth, this is approximately $3x10^{-7}$ cm/sec relative to a distance >> 20 AU.

According to GEQ the expansion does exist and is entirely due to the difference between equations (6) and (7) that can be measured under certain conditions. One possibility is the decay of high-orbit satellites, but that approach is confounded by other factors such as the presence of small quantities of gases and solar pressures, so an alternate earth-bound approach is suggested here.

Specifically, a vertical column made up of ridged but compressible foam that is measured for height-changes in time as described below. The reason for using a compressible material is because in theory the effects of the predicted expansion process is greater at the bottom of the column than it is at the top so some means for absorbing the internal stresses occasioned by the expansion process is required. As such, the magnitudes of the height changes listed below in Table II are theoretical maximums and the expectation is that the actual measured values will be somewhat less but still observable. As indicated the tests require observation over extended times, how long before the changes can be observed depends on the heights of the columns and the ability of the compressible foam used to absorb internal stresses. _(One way to calibrate the properties of the column's ability to absorb internal stresses is to mechanically raise the column at its base and measure the displacement at its top prior to the suggested measurement protocol It may be appropriate to add a small mass to the top of the column to obtain reproducible, accurate results)._

The essence of the new test is to measure the height of the "rigid" column for an extended period of time using a Michelson Interferometer as indicated in Figure 23 with a reflective termination at the top of the column such as to cause the laser to develop standing waves. By measuring the position of a node of the standing wave any change in height will be observed as a change in the node location. Using the movable mirror to return the detected node to its original location will provide the magnitude of any change.

While the predicted change per unit time is small, the integrated changes over time are well within experimental capability as indicated in Table II. If as expected the column height decreases at a rate commensurate with the distortion relation of equation 11 the expansion hypothesis will be supported. If not, it will be falsified.

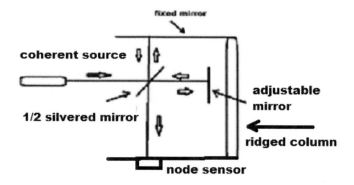

Michelson Interferometer

FIGURE 23

As an example, with reference to Figure 23, if I use a Helium-Silver laser as the coherent source, the distance between nodes is 224.3 x10^{-9} meters (224.3 nanometers). The predicted change in node location as a function of time for various heights of the "rigid" column, based on equation 11, are listed in Table II and should be measurable to adequate accuracy if they exist.

Derivation of distortion parameter

According to GEQ theory there is a measurable correction in acceleration due to the difference between equations 6 and 7 that can be expressed as:

$$acceleraton = \frac{GM}{R^2}\left(\frac{t_{eq}}{t} - 1\right) \quad \text{Where:}$$

$$t_{eqe} = \frac{R}{C}(1 - \frac{R_{miin}}{R})^{1/2} + \left(\frac{R_{miin}}{C}\right) ln\ [\left(\frac{R}{R_{min}}\right)^{\frac{1}{2}} + \left(\frac{R-R_{min}}{R_{min}}\right)^{\frac{1}{2}}\ ]$$

$$\text{and} \qquad t = \frac{R}{C} \qquad \text{or} \ \ R = C * t$$

Expressing this as a constant extra velocity at some distance from the center of mass of a large body, one has:

$$V = \frac{GM}{R^2}\left(t_{eqe} - t\right)$$

Solving this for velocity at the surface of the earth ($R_e = 6.371 \times 10^8$ cm), one has:

$$Ve = \frac{GM}{R^2}\left(t_{eqe} - t_e\right) = 3.02 \times 10^{-7} cm/sec$$

Then, for some higher co-linear distance (R_x) above the earth, the predicted change in velocity is given by:

$$\Delta V = (\ Ve - Vx) = 3.02 \times 10^{-7} cm/sec - \frac{GM}{(R_x)^2}\left(t_{eqx} - t_x\right) \tag{11}$$

And $\Delta R = \Delta V * t$ where "t" is the time interval of the measurement

PREDICTED DISTORTION-VELOCITIES AND CHANGES IN ELEVATION AND NODE POSITIONS AS FUNCTION OF TIME

ELEVATION DIFFERENCE METERS	DISTORTION VELOCITY CM/SEC	CHANGE IN ELEVATION IN 15 Days nanometers	CHANGE IN ELEVATION IN 30 days nanometers	CHANGE IN ELEVATION IN 90 days nanometers	CHANGE IN ELEVATION IN 180 days nanometers	CHANGE IN ELEVATION IN 360 day nanometers
1	9.49×10^{-14}	1.24	2.48	7.44	14.8	29.6
2	1.86×10^{-13}	2.45	4.9	14.7	29.4	58.8
5	4.63×10^{-13}	6.1	12.2	36.6	73.2	146.4
20	1.85×10^{-12}	24	48	144	288	576
50	4.6×10^{-12}	61	122	366	732	1,464

For greater elevations, the predicted values for growth for heights varying from 200 meters to 10,000 meters are shown in Table III.

TABLE III

PREDICTED DISTORTION-VELOCITIES AND CHANGES IN ELEVATION AS FUNCTION OF TIME

ELEVATION DIFFERENCE METERS	DISTORTION VELOCITY CM/SEC	CHANGE IN ELEVATION IN 15 Days mm	CHANGE IN ELEVATION IN 30 days mm	CHANGE IN ELEVATION IN 90 days mm	CHANGE IN ELEVATION IN 180 days mm	CHANGE IN ELEVATION IN 360 day mm
200	1.86×10^{-11}	2.4×10^{-4}	4.9×10^{-4}	14.7×10^{-4}	29.4×10^{-4}	58.8×10^{-4}
500	4.63×10^{-11}	6.1×10^{-4}	12.2×10^{-4}	36.4×10^{-4}	72×10^{-4}	144×10^{-4}
1000	9.25×10^{-11}	12.2×10^{-4}	24.4×10^{-4}	74.4×10^{-4}	148×10^{-4}	296×10^{-4}
2000	18.6×10^{-11}	24.52×10^{-4}	49×10^{-4}	147×10^{-4}	294×10^{-4}	588×10^{-4}
5000	4.6×10^{-10}	6.1×10^{-3}	12.2×10^{-3}	36×10^{-3}	72×10^{-3}	144×10^{-3}
7000	6.47×10^{-10}	8.4×10^{-3}	16.8×10^{-3}	$50,4 \times 10^{-3}$	101×10^{-3}	202×10^{-3}
8000	7.4×10^{-10}	9.8×10^{-3}	19.6×10^{-3}	58.8×10^{-3}	117.6×10^{-3}	235×10^{-3}
9000	8.32×10^{-10}	10.1×10^{-3}	20.2×10^{-3}	60.6×10^{-3}	120×10^{-3}	240×10^{-3}
10000	9.24×10^{-10}	12.2×10^{-3}	24.4×10^{-3}	73.2×10^{-3}	146.4×10^{-3}	292×10^{-3}

There are several observations to be made about these data before moving on to the question of the overall growth rate of the earth.

The first is to note that the values obtained depend on the distance between the earth's surface and the measurement height. This of course is very different from measurements made in inertial frames where the observer and the observed are at rest relative to each other and the result only depends on V_1. That is, when it is just the acceleration of the earth and it doesn't matter where the observer is located. Because the measurements are made in non-inertial frames, the observers are not at rest relative to each other and the values they measure in each case are as defined by equation 11, increasing monotonically as the magnitude of V_2 decreases with distance. In other words the distortion velocity increases with measurement distance as indicated in column 2 of both Tables.

Accordingly, at least in my understanding of the issue, previous data does not provide an answer to what is meant by "expansion of the earth as a whole" because they do not supply a reference distance from which the data is accumulated. I can calculate backwards and estimate an elevation that fits their mean value and it turns out to be about 20 km. Since the highest point on earth is Mt. Everest (8.9 km) and the lowest satellite for geometric measurements has been about 160 km (Tsubame- Japanese for barn swallow), it is unclear how they arrived at this value in terms of direct measurements.

Secondly, since Wenbin Shen et al. do not provide details of their calculations it is difficult to compare their results with the above calibration data because the uncertainty is so large (0.35 +/- 0.47 mm/year). However if they sorted their data by elevation it might be possible to compare their data to Tables II and III.

What is of prime interest is that regardless of the accuracy of their data, it appears that the notion the earth is expanding in time is a very strong belief in the relevant research community and this tends to support the basic teaching of GEQ.

Let me now address the overall growth of earth in the context of Tables II and III by considering what the distortion velocity would be for observations made from different elevations above the earth.

If the measurement were made from a satellite at say 500 km, the distortion velocity would be about 4.1×10^{-8} cm/sec and the increase in radius of the earth measured would be about **6.5 mm/year**.

If made from 800 km, the distortion velocity would be larger (6.2×10^{-8}cm/sec) and the increase in radius measured would be about **9.8 mm/year**.

According to the data reported in the Wenbin paper the value from space geodetic measurements is 3.2 mm/year for ocean rise and this includes 1.8 mm due to water expansion because of ice melting, so the solid earth growth is estimated to be **about 1.4 mm/year**, vastly less than the above values taken, respectively, at 500km and 800km.

To match the space geodetic values, according to the GEQ model, the measurements would have to be made from an altitude of about 100 km (distortion velocity 9.08×10^{-9}cm/sec) yielding a radius increase of about **1.4 mm/year.**

The difficulty is that over the twenty years the space geodetic data has accumulated there have not been any satellites at 100 km so here is a large unexplained difference between the GEQ claim and existing data. The question is, why?

Either the GEQ reasoning is wrong or the existing data fails to report the effective altitude of their measurements. I notice that in all data reported that no mention is ever made that the measured expansion depends on altitude of the observer. So either I have it wrong or the community has it wrong. If I were an unbiased observer to this issue, I would bet on the community. But I'm not an unbiased observer and I bet on GEQ!

The remainder of the chapter assumes the expansion exists and predicts what the parameters would be if the distance between the tip of the eastern nose of Brazil to the western boundary of Gabon in Africa, at present bit more than 3000 miles, was reduced to 3 miles over a period of 175 million years. This would correspond to the apparent change based on fossil data and infers the radius of the earth has reduced from its present value of about 6.4 x 108 cm to 6.4 x 105 cm. I am comfortable in using these values because at best my results are going to be just an approximation.

These values are compared to the "calibration values" from Table II where it is assumed the anomalous acceleration of the earth causing the growth is approximately 1.4×10^{-5} cm/sec^2.

What becomes apparent is that the definition of what is meant by time interval in the past were not the same as today, so the "actual" time lapse in years lies somewhere between 175 million years and 147 million years, depending on the definition used. Given this correction there is satisfactory agreement between calibration and predicted values.

One caveat: Note that this change in size by a factor of 103 does not imply the earth's radius was actually only 6.4 x 103 cm 175 million years ago. It refers to what a hypothetical observer "outside the universe" would see. We would experience no significant change in dimensions because we and our measuring sticks would likewise change in size. It only refers to the *distortion velocity term* not the actual size. It is an artifice useful for the coming calculation. Its meaning is a measure of how much the expansion process deviates from the *ratio of sizes remaining constant over time.*

A big question is, "What was the mass of the earth 175 million years ago?" The only source for the expansion must come from the earth's mass so presumably the mass must have been greater in the past: the issue is, how much greater? To calculate this as a first approximation let me consider the work done to move the center of mass of the earth resting against an "immovable wall" from its present radius (6.4×10^8 cm) to infinity. One finds this amount to about 4×10^{18} grams while the earth has about 6×10^{27} grams. Given these results it is reasonable to assume the mass is a constant.

Now what I need is the mean value for the *distortion velocity* of the earth when its radius went from 6.4×10^5 cm to the earth's present size of 6.4×10^8cm. That value, adjusted by the ratio of length to the earth's circumference, multiplied by 175 million years provides me with the net change in geometry for Pangea over that time interval.

Let me first consider what magnitude of average distortion velocity would result in a change in distance of a factor of 10^3 after 175 million years. Using:

175 million years $=5.5 \times 10^{15}$ sec:3 miles $=4.83 \times 10^5$ cm.

3000 miles$=4.83 \times 10^8$ cm. .; Radius$=6.4 \times 10^8$ cm

I obtain:

$$6.4 \text{cmx} \frac{(10^8 - 10^5)}{5.5 \times 10^{15} \text{sec}} = 8.73 \times 10^{-8} \text{cm/sec}$$

Thus the average distortion velocity for the distance between S. America and Africa is 8.73×10^{-8} cm/sec according to this approach.

It is most convenient to solve the remaining issue of average distortion expansion velocity of the earth's radius increasing by a factor of 10^3 by plotting the function of radius over time. Figure 24 shows the behavior when the radius goes from 6.4×10^5 to 6.4×10^8cm. with a mean value of 3.2×10^8 cm. f(R) (is equation (6) multiplied by "C":

$$f(r) = (r)(1 - \frac{0.8879496}{r})^{\frac{1}{2}} + 0.887496$$

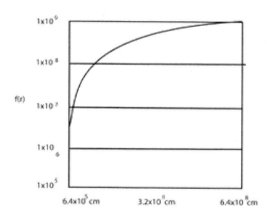

Value of f(r) over range of 10^5 cm to 10^8 cm

FIGURE 24

The calculated mean distortion velocity of the earth's radius from our perspective is given by:

$$\frac{G \cdot Me}{Rez^2} \cdot (teqz - tez) - \frac{G \cdot Me}{Re^2} \cdot (teqe - te) = 8.662532 \cdot 10^{-7} \text{ cm} \cdot \text{sec}^{-1}$$

Finally, the length of the present distance of the surface segment of interest is 3000 miles and that is approximately 0.12 times the expansion of the circumference (3000miles/25000miles) referenced to the change in radius, so the calculated value for the mean distortion value of :

$$.12 \cdot \left(8.662532 \cdot 10^{-7} \text{ cm} \cdot \text{sec}^{-1} \right) = 1.039504 \cdot 10^{-7} \cdot \text{cm} \cdot \text{sec}^{-1}.$$

As compared to 8.73×10^{-8} cm/sec as calculated from:

$$8.73 \cdot 10^{-8} \frac{\text{cm}}{\text{sec}} \cdot \left(5.5 \cdot 10^{15} \text{ sec} \right) = 4.8015 \cdot 10^8 \text{ cm}$$

The reason for this difference is because the first calculation gives the *mean time* in years, while the second calculation provides years in our timescale which is considered to be time invariant. Examining Figure 24 clearly indicates that time-intervals in the past were

$$\left(1.039504 \cdot 10^{-7} \text{ cm} \cdot \text{sec}^{-1} \right) \cdot 4.62 \ 10^{15} \text{ sec} = 4.802508 \cdot 10^8 \text{ cm}$$

And find the *mean time* is 4.62×10^{15} sec, or 146.6 million years.

SUMMARY OF CHAPTER 12

When NASA launched its Pioneer series of deep space probes in the early 1970s, for the first time, sufficient guidance and tracking technology existed to enable very precise measurements of distances, velocities, and timings of vehicle parameters. These technologies included both on-board equipment and ground- based monitoring equipment, particularly the Deep Space Network (DSN) with three sites located around the earth, so a continuous stream of vehicle behavior was available as the earth rotated on its axis and travelled around the Sun. The data eventually collected was for the two successful vehicles, Pioneer 10 and Pioneer 11.

When the vehicles were about 20 AU from the Sun (1 AU is the average earth-Sun distance and often used as a metric for astronomical distances), the measurement data began to show an anomaly. The anomaly appeared as an extra force from the direction of the Sun not accounted for by known physics theory. As the experiment continued it became evident that there were three anomalies, not one, which appeared to depend on the location of the earth relative to the Sun and the direction of the vehicle relative to the earth based DSN sites.

The magnitudes and timings of the three terms, expressed as accelerations, are important clues as to their causes. The main one focused on by the community had a magnitude of about 9×10^{-8} cm/sec2, secondly an almost sinusoidal diurnal term (synchronous with the earth's rotation around its axis) with a peak value about 10 times larger than the first term (100×10^{-8}

cm/sec2, and a third much smaller annual term of (0.2 x 10-8 cm/sec2) that occurred as a difference between the average measurement values when the earth was at its closest and furthest distance from the Sun. Understand that all these data were superposed on one another so the experimental team had to sort them out, not an easy task because there were many other "noise" sources also superposed in the data stream.

These anomalies led to extensive speculation as to the causes, including that they might herald "new physics," but no reasonable theories have emerged thus far.

In 2012, a new analysis headed by Slava G. Turyshev, a member of the original research group, re-examined the possibility that thermal radiation could explain the main anomalous term. This new analysis used additional data recovered from old files. While once again the thermal effects of the on-board nuclear power source were smaller than required to account for the anomalous term, and the uncertainty was large, the calculated values were within one sigma of the measured values. Dr. Turyshev did not address the remaining two terms and a one sigma level of certainty is well below the usual level required in physics, but his explanation has been generally accepted.

There is yet another anomaly NASA encountered in using the DSN, regarding what has come to be called the flyby anomaly. That often occurs on missions where the DSN cannot follow the vehicles in their early launch trajectories during "slingshot" around the earth to gain velocity for their missions. When vehicle tracking resumes, beginning trajectories deviate significantly from predicted courses. While it is not known that this effect is related to the Pioneer anomalies, I suspect it is, but I do not have sufficient data to model it, so it is not included in the following analysis.

The analysis by GEQ addresses all three anomalies, but since the effects are very complex, particularly the large nearly sinusoidal term, the results provided can only be considered approximate, although qualitatively I think they are convincing. I hope that NASA re-examines their data using the teachings provided herein. With their more powerful analytic tools, I suspect they would find the GEQ approach shows what they measured indeed represents "new physics."

Note that the Sun is the reference system used for all these measurements, and that all effects observed are local to the accelerated observer on the earth. They are not something the distant vehicle is doing. Those vehicles have no forces acting on them in the GEQ world.

For the main steady term of 10-8 cm/sec2, and the small annual term of 0.2 x 10-8 cm/sec2, the distances of the vehicles from the Sun used by NASA are given by equation (7); that is distance is equal to the length of time it takes a pulse of light to travel from the vehicle back to the earth, or back to the Sun.

But according to GEQ, the *actual distances* are given by equation (6) and those distances are always a bit shorter than the equation (7) values NASA uses for its calculations. Because of those differences, they measure slightly larger forces from the Sun than those they calculate, because Newton's gravity law (which is the law they use in their calculations) works that way. Shorter distances between a test mass and the source of a gravity field results in greater force.

For the small annual term, the difference between equations (6) and (7) changes a bit depending on the earth's distance to the Sun. When the earth is closest, equation (6) gets a bit smaller; when it is furthest from the Sun, it gets a bit larger. The net affect for the observer on the earth is that small annual term.

The analysis of the large diurnal term is also from the difference between equations (6) and (7), but in a much more complex way than for the other two terms. Roughly speaking its magnitude depends on the angle of any of the DSN sites to the vehicle as the earth rotates around its axis. Because of the very great distances involved, the data used by NASA usually involves more than one DSN site because tracking data is a complex process derived from radar signals transmitted to the vehicle, and other signals from the vehicle back to the DSN sites.

Round trip times are measured in tens of hours, so more than one site is involved in each calculation. According to the GEQ view, the large anomalous signal is from the effective growth of the earth during measurement intervals. Since that growth, if it exists, is always normal to the DSN site location, its magnitude changes with angle of transmission and reception, going to zero when an involved site is at 90 degrees to the vehicle.

As stated above the values developed in this analysis are reasonably close to the NASA values, but since I do not know all the statistical methods NASA used, I cannot directly compare my results to theirs, particularly in the case for the large diurnal term. In particular it is probable that their methods delete terms that have large deviations from the mean values (this strategy is common in statistical evaluations), but in this particular case those large differences are due to differences between angles of observations by the transmitting antenna and the receiving antenna. The consequence is that for the GEQ analysis where such data is included, the GEQ predicted value is much larger than the NASA one (about 170 cm/sec2 compared to about 110 cm/sec2). Other than this deviation the data are comparable.

CHAPTER 12

Anomalous Acceleration Of Pioneer 10 And 11

In 1972 NASA launched a deep space probe called Pioneer 10, the first man- made vehicle to pass out of our Solar System. NASA tracks any of these vehicles with a network of three ground-based stations that send and receive signals from the launched vehicles. Collectively called the Deep Space Network (DSN), stations are located at Goldstone, California, near Madrid, Spain and near Canberra, Australia. The sites are located at various longitudes allowing a continuous view of any probe as the Earth rotates on its axis. The "up" and "down" links are via microwave signals and the system further utilizes a complex system of computers to crunch and analyze data. From time to time the computer system generates control signals to adjust various aspects of the vehicle's orientation or other corrective maneuvers. For details of the space-shots, I refer readers to the extensive papers published on-line (arxiv) and in the American Physical Society Journal (APS) (Anderson, JD; et al., 2005) and (Nieto, MM, Anderson, JD. 2005) and G. Turyshev et. al (arxiv) 2012.

Beginning in 1980, when Pioneer 10 was about 20 AU from the Sun and the effects from the Sun's solar radiation had reduced to a small value, a systematic error in the plotted course was observed that implied an extra force coming from the direction of the Sun. Similar results were also obtained from data from the other deep space vehicles. This phenomenon became known as the anomalous acceleration. Here I will focus on just one of those vehicles, Pioneer 10.

In actuality there are three terms which are not modeled. The first, the most studied one, is a more or less constant extra acceleration aimed towards the Sun of average magnitude $(8.7 +/- 1.3) \times 10^{-8}$ cm/sec2. The second is a small annual periodic variation of magnitude $(0.215 +/- 0.022) \times 10^{-8}$ cm/sec2. The third is a much larger diurnal periodic term of magnitude $(109 +/- 7.9) \times 10^{-8}$ cm/sec2.

While NASA scientists are not sure if the periodic terms are associated with the main anomalous term, the analysis provided here claims they are related.

In the original paper, published by the research group headed by John D. Anderson, a very thorough analysis of the possible causes for the anomaly concluded that there was no likely explanation. Special attention was paid to the possibility that thermal radiation from the on-board nuclear power source (four plutonium-238 radioisotope thermoelectric generators) accounted for the anomaly, i.e., it was found that the external thermal radiation was too small to account for the extra acceleration.

In 2012 a new analysis headed by Slava G. Turyshev, a member of the original research group, re-examined the possibility that thermal radiation could explain the main anomalous term. This new analysis used additional data recovered from old files. While once again the thermal effects of the on-board nuclear power source were deemed smaller than required to account for the anomalous term, and the uncertainty was large, the calculated values were within 1 sigma of the measured values. This explanation was accepted by both the general physics and public communities. The so-called Anomalous Acceleration went away.

However, in this second analysis there were several findings that did not accord with the behavior of the anomaly. The most marked one was the major onset of the extra acceleration at 20 AU. The thermal radiation explanation implied it should have become evident at a much closer distance to the Sun with a magnitude *larger than* values at greater distances. Turyshev's explanation of this, supported by a large solar force in the *outward* direction was that perhaps there was an error in modeling. A second problem was that the one sigma level for data validity has about a 68% chance of repeated measurements falling within the range of the data. A 32% chance of falling outside that range is a far lower level of certainty than one would desire. A third difficulty, according to our analysis, is that the other two terms (annual and diurnal) are periodic and could not be due to thermal effects. They were most likely caused by the same factor that caused the steady-state term. In any event, it is probable that some of the anomalous acceleration *was* caused by thermal factors, but not all of it.

The model used in our analysis is only approximate. It treats the steady-state anomaly and the annual terms as one body of problems involving only the Sun, with an observer positioned at the earth's location. It treats the diurnal (daily) term as involving only the rotating earth, with observers located at each of the DSN locations. In all cases, the coordinate system used is centered on the center-of-mass of the Sun.

In contrast, the NASA solution uses parameterized post-Newtonian (PPN) approximations of gravitational theory and takes a coordinate system centered on the barycenter of the Solar System masses. Furthermore, in the actual NASA/DSN measurements all effects are included in the data stream. In this analysis my solutions are separated into three independent parts since the complex interactions among different parameters of the problem require computing capability, I do not have access to. Accordingly, my methods are crude compared to the NASA methodology. Despite this limitation, the results I obtain are respectably close to the actual observations. They show no anomalous acceleration below about 2 AU (presumably for the same reason as Turyshev's data).

In the first two analyses the assumption is made that the earth is absent and an observer is in Earth orbit around the Sun. This solution provides a magnitude for the steady term somewhat larger than the reported data (about 6% above the range reported by NASA), but shows the same overall pattern. It also provides an explanation for the annual term via a second method. The third analysis just uses the Earth in the model, considers locations of the three DSN sites from which measurements are made, and focuses only on the diurnal term. This analysis indicates that the observed value of the diurnal periodic term is a reasonable one in the context of the measurement techniques used.

Analysis of Steady-state Term: Figure 24 is a plot of the anomalous effect taken from the paper by Anderson et al, 2005.

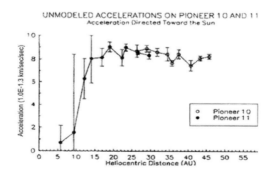

From: Study of the anomalous acceleration Pioneer 10 Anderson et al APS 2002

ANOMALOUS ACCELERATION PLOT

FIGURE 24

As seen in Figure 2, the extra acceleration first made its appearance in the vicinity of 5 AU, increasing to its average value around 20 AU. NASA assigns the magnitude as being between 7.31 and 10.07 x 10-8 cm/sec2 with a mean value of 8.7 x 10-8 cm/sec2.

In the method used for this first calculation of the anomalous term, as stated, I use a coordinate system centered on our Sun, consider only the mass of the Sun, and simply modify the Newtonian approach as follows: First, recalling equations 6 and 7 from Chapter 2:

Now, consider the velocity change for an observer 1 AU from the Sun (no earth present) where Newton's equation is taken with a plus sign. I can consider the acceleration a constant since, from our point of view, the Sun's size is time invariant.

For eq. 7 $\quad V_x = \dfrac{GM}{(R_{1AU})^2} t_x$ $\qquad$ For eq. 6 $\quad V_{eqx} = \dfrac{GM}{(R_{1AU})^2} t_{eqx}$

Evaluating this for the given conditions, I calculate the difference between the two as:

$$V_{eqx} - V_x = \dfrac{GM}{(R_{1AU})^2} \left(\dfrac{t_{eqx}}{t_x} - 1\right) t_x = 0.000053 \text{ cm/sec}$$

Dividing this result by t_x gives me the small extra local acceleration caused by the assumptions from the expansion hypothesis. In this case it is 10.6×10^{-8} cm/sec^2.

SUN $\qquad\qquad\qquad\qquad$ EARTH $\qquad\qquad\qquad\qquad\qquad$ PIONEER 10

1 AU

GEOMETRY OF OBSERVATION

FIGURE 25

What I am interested in is the difference between this value and the excess acceleration obtained in the same manner for the vehicle at various distances. A typical form, assuming the observer is always in-line with

Pioneer 10 (as shown in Figure 25), is given by: at >Table IV provides the calculated values for anomalous terms as a function of Pioneer 10's distance from the Sun with the observer 1 AU from the Sun as indicated in Figure 28. Since the function rises so rapidly, Table V supplies the same information for Pioneer distances 1.1 to 2 AU.

TABLE IV values of anomalous terms Pioneer 10

OBSERVER x 10^{-8} cm/sec^2	PIONEER AT > 1AU x 10^{-8} cm/sec^2	TOTAL x 10^{-8} cm/sec^2
1 AU (10.6)	5 AU (0.09)	10.5
1 AU (10.6)	10 AU (<0.001)	10.6
1 AU (10.6)	20 AU (<0.001)	10.6
1 AU (10.6)	30 AU (<0.001)	10.6
1 AU (10.6)	40 AU (<0.001)	10.6
1 AU (10.6)	50 AU (<0.001)	10.6
1 AU (10.6)	60 AU (<0.001)	10.6
1 AU (10.6)	70 AU (<0.001)	10.6

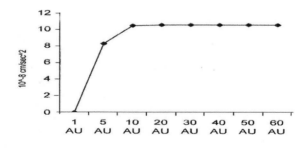

PLOT OF TABLE IV

FIGURE 26

TABLE V values of anomalous terms Pioneer 10

OBSERVER x 10^{-8} cm/sec^2	PIONEER AT > 1AU x 10^{-8} cm/sec^2	TOTAL x 10^{-8} cm/sec^2
1 AU (10.6)	1.1 AU (8.0)	2.5
1 AU (10.6)	1.2 AU (6.2)	4.3
1 AU (10.6)	1.3 AU (4.9)	5.6
1 AU (10.6)	1.4 AU (4.0)	6.5
1 AU (10.6)	1.5 AU (3.2)	7.3
1 AU (10.6)	1.6 AU (2.7)	7.8
1 AU (10.6)	1.7 AU (2.2)	8.3
1 AU (10.6)	1.8 AU (1.9)	8.6
1 AU (10.6)	1.9 AU (1.6)	8.9
1 AU (10.6)	2.0 AU (1.4)	9.1

TABLE V values of anomalous terms Pioneer 10

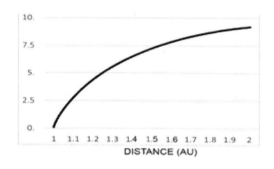

PLOT OF TABLE V

FIGURE 27

Analysis of Annual Term The orbit of the Earth around the Sun is not quite circular, but elliptical. If instead of taking the observer's distance as fixed at 1 AU from the Sun, I consider its distances at aphelion and perihelion, solutions to the equation give me slightly different magnitudes for the anomalous term:

At aphelion	10.44×10^{-8} cm/sec^2
At perihelion	10.78×10^{-8} cm/sec^2

Taking half of this difference as the magnitude of a yearly periodic term, one has 0.17×10^{-8} cm/sec^2 as a possible source for the annual term. While this is a bit smaller than the magnitude of the measured term $(0.215 +/- 0.022) \times 10^{-8}$ cm/sec^2, it could be that it combines with residuals of some error function or that differences in the models used account for it. Also, according to this analysis, maximum peak occurs near perihelion (early January), while the NASA analysis indicated the maximum peak at conjunction about two weeks earlier than perihelion. Again, this difference may be due to differences in the modeling.

Analysis of Diurnal Term The data sorting for this term is significantly more difficult to analyze than the other terms and indicates something about the difficulties NASA incurs in all the measurements. Understand that all the data they collect is commingled, so the issues raised for this analysis are also required for the other two effects.

If the view of the probes were directly in line with the observer the expansion calculation would indicate a magnitude for the diurnal term more than seven times larger than the reported value of approximately $110 \times 10-8$ cm/sec2. This follows directly from equation (6), used the same way as used for the Sun where the Earth was considered absent. However, since the DSN stations are never in-line with the probes, the angles of observation lower the measured values significantly. While the presence of the Sun also influences these values, that effect is exceedingly small compared to the earth, so I can ignore that factor in this analysis.

The three DSN sites are distributed around the globe covering a full 360o view. The coordinates of these sites are: Goldstone, Ca. (latitude: N 35o,

120

longitude: 117o west), Canberra, Australia (latitude: S 35o, Longitude: 149o east) and Madrid, Spain (latitude: N 40o, longitude: 4o west).

All three sites use high-gain steerable parabolic antennae arrays aimed towards the target vehicle both for transmission and reception of signals. Since the "up-link" and "down-link" signals overlap in time, the carrier frequency of the downlink is shifted to avoid interference or confusion.

The frequency shift is done in such a manner as to retain coherency, so phase information is not lost (phase information of the signals carries most of the data used for analysis on vehicle parameters). Since the sites are separated by many kilometers it is necessary that they be time coordinated. A universal timing system is used, established from the Global Positioning System (GPS), with very accurate clocks at each site to supplement the timing data. Part of the data stream in every transmission contains a digitally encoded time tag. All data received from measurements is recorded and crunched later to obtain target vehicle data (distance, angles, velocity, and acceleration). Note that the elapsed round-trip light time between transmission and reception is measured in hours, ranging from about 5.5 hours (when the vehicle is at 20 AU) up to about 16.5 hours (when the vehicle is at 60 AU). Because of these large time-lags and the rotation of the Earth any single piece of information uses multiple antennae even though it is the signal from one antenna that is transmitted. The reception of that given signal will generally be by one or more different antennae.

Before looking at the interaction of the different antennae in measurements, let me first establish that signals from the expansion hypothesis do indeed look like periodic terms with periods equal to the Earth's rotation. Consider a situation with three antennae mounted on a non-rotating Earth, where the antennae are spaced 900 apart: the center one directly in-line with a distant target, the two outermost antennae at the east and west horizons. Since the hypothesized expansion is normal to the Earth's surface at every point the two outermost antennae will have no evidence of any expansion signal for their transmissions and receptions, while the center antenna will respond with the maximum amplitude of that expansion whatever value that might be. If I add additional antennae, spaced between these three, the magnitudes of the received signals will trace out a function appearing approximately as a sine wave going from a minimum, through a maximum and back to a minimum. Hence, the response from the collective antennae will appear as though there was a sinusoidal term emanating from the Earth. In the dynamic situation, where the Earth is rotating, analysis of the complex signal received will label it with a magnitude equal to one half of the measured peak-to-peak amplitude.

Having said the above, the real-world measurement routine requires corrections to the received data for factors related to the geographic location of each member of the antennae complex and other factors including, corrections for changes in phase introduced by the antennae when their elements are adjusted to aim them (if such is the case), corrections relating to the non-rigidity of the Earth, and corrections for the Earth's motion around its own axis and around the Sun.

Since the nominal assumption is that the Earth's size is time invariant, there is no correction made for any growth factor such as implied by the expansion hypothesis. While the NASA measurements indicate the diurnal term has a constant magnitude lying between 101 x 10-8 cm/sec2 and 117 x 10-8 cm/sec2, the analysis here predicts a mean magnitude on the order of 170 x 10-8 cm/sec2.

However, what is unambiguous about the prediction is that a diurnal term should be present with minima always occurring when a transmitting antenna is just rotating into or out of view of the distant Pioneer 10.

Let me take as a test situation a time when Pioneer 10 is 40 AU from the Sun with a round-trip light time of just over 11 hours for a signal traveling from the Earth to the target vehicle and back. What I will focus on is the peak maximum of the sinusoidal term, which will occur when any of the 3 antennae have longitudes in-line with the Pioneer 10. The assumption is that the stored data from each of the sites will be combined during later analysis to extract the value of the measured diurnal term.

At 40 AU, the location of Pioneer 10 is given in ecliptic coordinates as longitude = SE 70.9o and latitude = SE 3.1o (see **.http://cohoweb.gsfc.nasa. gov/ helios/**). Converting these to equatorial coordinates, one has, latitude/ declination = 25.15o S and longitude/right ascension = 68.84o. The location of Pioneer 10 expressed in equatorial coordinates, along with the DSN station locations as given above, now allows me to calculate the effective measurement results considering the decreases in magnitudes caused by the angular views of the hypothesized expansion terms.

The angles are imaged in Figures 28a and 28b for the situation where the transmission takes place when the longitude of the Goldstone station is exactly aligned with Pioneer 10 at 68.84o. Figure 28a is the view looking down on the Earth from above the North Pole. Figure 28b is the view looking along the equator.

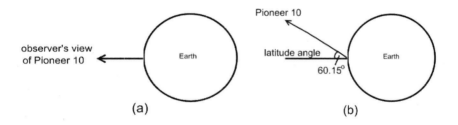

(a)   (b)

Viewed from North Pole   Viewed along equator

Views of angular relationship between measurement sites and The Pioneer 10 using Goldstone, CA. as the transmitter site.

FIGURE 28

Since the proposed expansion of the Earth takes place normal to the Earth's surface at the transmitter site location, during transmission the original magnitude of the diurnal term is reduced by the single angle shown in Figures 30b, so the transmitted magnitude is given by:

$$(1422 \times 10^{-8} \text{ cm/sec}^2) \cos(60.15°) = 707 \times 10^{-8} \text{ cm/sec}^2$$

Figures 29a and 29b depict the state of the receiver sites some 11 hours later, when the transmitted signal has completed its round trip from the Pioneer 10 back to the Earth. During this time, the Earth has rotated about 166o, so Canberra is the receiving antenna.

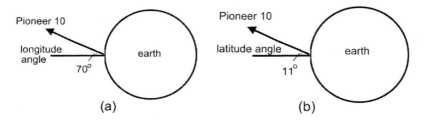

<table>
<tr><td align="center">(a)</td><td align="center">(b)</td></tr>
<tr><td align="center">Viewed from North Pole</td><td align="center">Viewed along equator</td></tr>
</table>

Views of angular relationship between measurement sites and Pioneer 10 using Canberra as the receiver site.

FIGURE 29

On reception by the antenna at Canberra another reduction is incurred due to the two angles shown in Figures 29a and 29b, so the final peak-to-peak amplitude is 238 x 10-8 cm/sec2.

Taking half of this value gives me about 119 x 10-8 cm/sec2 as the magnitude for the diurnal term.

If I repeat this method for each of the other two sites, in each case starting with the transmitting site facing Pioneer 10, I obtain a value for the Canberra/Madrid pair of about 285 x 10-8 cm/sec2.

When the Madrid antenna is at the same longitude as Pioneer 10 there are two receiving stations aligned after 11 hours to receive signals (Goldstone and Canberra). So here the results are: for the Madrid/Goldstone, 90 x 10-8 cm/sec2; and for Madrid/Canberra, 209 x 10-8 cm/sec2. If these last two are averaged the result is 150 x 10-8 cm/sec2.

For final analysis of these data points, presumably, the data is time shifted for each site so that they coincide to provide one day's (24 hours) observation. The values are smoothed and averaged to provide a best-fit result. In that case one would obtain an average for the peaks of this day of about 185 x 10-8 cm/sec2.

After 24 hours, the Goldstone station would be back in alignment with Pioneer 10. So, except for motion of Pioneer 10 (which would be small), the same data point collection would occur for the next day.

It is clear from the above that the light-time travel between the stations and Pioneer 10 plays an important role in what these magnitudes will be. This is because it determines what the observation angles will be for the receiving sites. The example given in the paper by Anderson et al. (2002) uses a distance of 66 AU with a 2-way light travel time of 18.3 hours (274.5o rotation), so it is worthwhile to examine the scenario for that distance as well. I obtain:

For Goldstone/Madrid, about 130 x 10-8 cm/sec2; for Canberra/Madrid about 300 cm/sec2; and for Madrid/Canberra, about 76 cm/sec2. The average of these, after time shifting for each site, is about 168 x 10-8m/sec2.

The conclusion from the above is that while the expansion hypothesis demands that the diurnal term be present, and that its form is a sinusoid with minima occurring just as the receiver sites come to view the distant Pioneer 10, the constant magnitude of about 100 cm/sec2 indicated by the NASA measurements is considerably less than that predicted by the model I use. It may well be that the differences in the models and the effects of the intervening values between the maxima and the minima change the best-fit values for the maxima. However, this is by no means obvious. Also, it may be true that other corrections NASA makes to account for the Earth's rotation reduce the value somewhat (including the possibility that they reject some data lying well away from mean values).

In summary, the predicted term for the anomalous acceleration is on the order of: 10.6 x 10-8 cm/sec2 for the steady term, 0.165 x 10-8 cm/sec2 for the annual term and on the order of 170 x 10-8 cm/sec2 for the diurnal term (given the model used and the measurement methods assumed). None of these values account for all measurement details. They should, however, serve as some degree of evidence for the validity of the expansion hypothesis as an explanation for gravity. For comparison, these values compare to measured values of 8 x 10-8 cm/sec2 to 10 x 10-8 cm/sec2 for the steady term, about 0.215 x 10-8 cm/sec2 for the annual term, and 101 x 10-8 cm/sec2 to 117 x 10-8 cm/sec2 for the annual term.

In as much as I criticized the Turyshev solution for its statistical weakness, it is only fair that I apply the same criteria to my solutions. While Turyshev's data is within one sigma agreement with the steady-state term, my solution is greater than one sigma away for that term and likewise greater than one sigma away for both the annual and diurnal terms. Considering the approximate nature of my analysis, that is not surprising. What is more important is that I can address all three anomalies. I would hope that NASA would undertake a new analysis of their data including the methods/teachings included in this above assumption.

SUMMARY OF CHAPTER 13

The essential purpose of this chapter is to point out several differences between the conventional thinking about cosmology and the GEQ viewpoint. The differences play a large role in interpretations in chapters 14 and 15.

One difference is the focus GEQ places on system dynamics when orbital velocities *increase* with distance rather than decreasing according to Newton's inverse law as in the behavior of planetary circular velocities (planets further from the Sun circle at slower rates). This inverse law behavior is in contrast to the velocity distribution *within* a rigid body where it is known that any such body that has angular momentum is characterized by a monotonic *increase* of circular velocity with increasing distance from the center of the body.

Based on this observation GEQ postulates that any system composed of discrete elements exhibiting a monotonic increase of circular velocity (a spiral galaxy, for example, made up of groups of stars) should be considered as a *rigid* body in an extended sense, implying that at some time in the past it must have been a single entity. The consequence of this view is that the rotation properties of any ensemble of stars retains angular momentum from its previous incarnation and this must be considered in understanding the dynamics of its present behavior. As applied in chapter 14 it does away with dark matter at least as far as spiral galaxies are concerned.

The same considerations play a large role in chapter 15, where the issue is what we on the earth see of distant galaxy motion where once again radial velocities are observed to increase monotonically with distance. However since the distances are much greater than the intra-galaxy behavior other factors come into play as well that result in even greater differences between the GEQ and the conventional view.

The most obvious of these, addressed in chapter 15, is that while the conventional view is that space stretches at speeds faster than light, but that the stretching only occurs at galactic sizes and above, the GEQ view is that all "stretching" or expanding occurs within matter itself, so it applies down to the smallest level.

One consequence of this difference is that all conventional graphic plots of distances between galaxies and the like are plotted as exceeding light velocity, while all GEQ plots use the dictate from Special Relativity that nothing can go faster than light. Unlike the situation in chapter 14 where all velocities are much less than the speed of light, distant galaxies have speeds great enough so that GEQ velocity plots have to be converted into nonlinear forms, where very distant galaxies appear to be traveling slower than predicted by the conventional view.

This will be made clearer in chapter 15.

The conventional view leads to the idea that relatively nearby galaxies are being accelerated (read, dark energy), while GEQ says no, simply a difference in interpretation of the data based on a metrics issue. These metric issues also involve the difference in values of equations 6 and 7, the derivation of the GEQ expansion process introduced in chapter 2. The details of how that game plays out appears in chapter

CHAPTER 13

Cosmological Models

The teachings of GEQ allows a distinctly different interpretation for distribution of radial and/or tangential velocities of observed masses than conventional Newtonian physics, the difference becoming centrally important for cosmological models. To understand why this difference arises it is necessary to examine equation (3) from the expansion derivation presented in chapter 2. For convenience let me present the entire equation sequence here .

DERIVATION OF EXPANSION EQUATION

I start by considering an arbitrary sphere that expands in accordance with Newton's gravity law with a plus sign:

$$Acceleration = \frac{d^2R}{dt^2} = +\frac{GM}{R^2} \tag{1}$$

I note: $d(\frac{dR}{dt})^2 = 2\left(\frac{dR}{dt}\right)\left(\frac{d^2R}{dt^2}\right) dt = 2\left(\frac{d^2R}{dt^2}\right) dR = (-\frac{2GM}{R})$ \hfill (2)

Assuming that the initial velocity is zero and (from Special Relativity) the maximum velocity reached is the velocity of light (c), I obtain:

$$Velocity = \frac{dr}{dt} = (C^2 - \frac{2GM}{R})^{1/2} \tag{3}$$

If I define the minimum size the sphere had at the beginning of the process (t =0) as (Rmin),

I can write an additional relationship that is identical to the Schwarzschild limit.

$$Schwarzschild\,limit = R_{min} = \frac{2GM}{C^2} \tag{4}$$

Let me invert the velocity term to the form dt/dR :

$$\frac{dt}{dR} = \frac{1}{(C^2 - \frac{2GM}{R})^{1/2}} \tag{5}$$

Integrate to obtain the relation between time (t) and size (R). Solve for the constant of integration, rearrange, and replace all terms of form (2GM/ C²) with Rmin. Properly combine terms and I finally obtain:

$$t_{eq} = \frac{R}{C}(1 - \frac{R_{min}}{R})^{1/2} + \left(\frac{R_{min}}{C}\right) ln[(\frac{R}{R_{min}})^{\frac{1}{2}} + (\frac{R-R_{min}}{R_{min}})^{\frac{1}{2}}] \tag{6}$$

In equation (6), I label the left-hand time term as "teq" for convenience later in this development. After a long time passes, the first term dominates and the equation converges to:

$$t = \frac{R}{C} \qquad or \quad R = C*t \tag{7}$$

Notice in particular that equation (3) says that the radial expansion velocity of a so-called rigid sphere increases monotonically from zero to c as the size R increases. This is the reverse of what Newton's gravity law tells us for any simple experiment involving either radial velocities, or by extension, since rotational speeds are directly related to radial velocities.

On the other hand, while GEQ predicts the same results for the same experiments performed under the same conditions, it adds another prediction for any object *before* it is thrown and is left resting on the earth. Specifically, it claims the object is not at rest, but that the earth is increasing in size and consequently the object is accelerating away from the center of the earth! This prediction is in accordance with equation (3) and the manner in which, according to GEQ, rigid bodies behave.

The important question is, for any given process where I want to analyze the causal details, should I start with a model based on the way our Solar System behaves or on equation (3)? Evidently the answer must be driven by whether the physical process being studied is best described as analogous to our Solar System with discrete bodies separated by space and interacting via gravitation, *or* as, in some sense, rigid bodies expanding in accordance with equation (3). In the former case, I can model the gravity field intensity decreasing with distance from some central mass, or in the latter case, increasing .It seems clear the choice should be driven by the observed behavior of the system being studied prior to any complete understanding of underlying details of forces and whatever else is involved causing the behavior.

The next two chapters discuss the behaviors of the tangential velocity distribution of elements in spiral galaxies (in chapter 14) and the velocity distribution of distant galaxies as viewed from the earth (in chapter 15). As you will see, both of these processes behave more like rigid bodies in the sense that velocities *increase* with distance from a reference center, rather than *decrease* with distance (at least until they get up to a significant percent of the velocity of light).

It certainly is no great surprise that rigid bodies behave this way; it is in keeping with the understanding of standard Newtonian and General Relativity theory, as is evident if one considers the circular velocity of some object in the earth at various depths. One obtains the plot shown in Figure 30, where the circular velocity increases linearly with distance from the center.

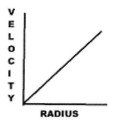

FIGURE 30

What is of great surprise is that these distributed, apparently separated bodies should act as though they were rigid bodies in the broader sense. There is nothing in either standard Newtonian or General Relativity theory that explains why this should be expected. That includes Alan Guth's notion of "spatial stretching," which you will read about later, because he explicitly excludes that the stretching or expansion apply to local groups such as our solar system or within galaxies.

What GEQ suggests is of greatest interest in addressing this puzzle are two factors: (1) That masses in orbit around any central mass _might possibly_ _be_ considered as *part of a rigid body in the broadest sense,* and (2) that the difference between equations 6 and 7, *adds an unexpected variable in any dynamic calculations of gravitational behavior in spatially distributed systems.*

Let me start by considering the first of these issues. It is evident that any object in orbit around the earth obeys a very different law than some object that is part of a rigid body in either the restricted sense or the broader sense. In particular the rotational law for test masses in orbits is given by the well-known equation:

Tangential speed = $(GM/R)1/2$ (12)

That is, the tangential speed to obtain a stable orbit around some central mass, gets less as the distance R increases.

In contrast, the law determining the conditions for the mass to retain its distance from the center of the large body, is entirely dependent on the angular momentum of the large body and the object's distance from the body's center. Let me elaborate on this a bit.

All large bodies (planets, stars, galaxies) possess angular momentum. Why this is true is not known, but the usual assumption is that at the time these objects were formed, the particles that accreted in their formations possessed rotations and passed that energy to the end product. Be that as it may, for any solid, truly rigid sphere, one understands the rotation speed of any sample taken at some arbitrary radius of the sphere is simply:

$$\text{Tangential speed } V_x = R_x/R_{max}(V_{max}) \tag{12}$$

That is there is a straight line-up of speeds at depth as indicated in Figure 32.

This is very different than equation 12 and due to very different considerations: simply, that the surrounding mass carries it along.

But in the case of a rigid body in the broader sense no such physical contact exists so apparently what must be true is that at some time in the past, they *had been in physical contact and even after they separated each* "hunk" retained its angular momentum. This gives rise to the monotonic increase in rotation speed as they get further from the reference center *until they reach velocities such that Einstein's speed limit comes into play.*

But why should they separate? And this gets me to factor (2). Unlike Guth's notion of spatial stretching or expansion the GEQ view is that it is above all else resident in the behavior of mass, so it has a definite effect locally. One of those effects arises from equation 3.

Any two elements of the mass where one is a bit farther out from the center, will have a slight differential velocity between them, the farther-out one going a bit faster, so *gaps may develop throughout the distribution of elements making up the total body.*

Given this, it is not too difficult to understand that as time goes on, what started out as a single connected entity subdivides into smaller entities retaining both angular momentum (speed of rotation) and radial velocities inherited from the original whole and characterized by their locations.

To go through the details of this process is well beyond my capabilities and the immediate needs of this book, so I will just assume it is a viable explanation of how planets, galaxies, and what else have you, form; in which velocities increase monotonically with distance.

Let me take these ideas and see how they apply to chapters 14 and 15. When you read each of them you will see the applications in much greater detail.

Chapter 14 focuses on why the rotation curves in spiral galaxies do not correspond to the predicted behaviors indicated by the amount of matter that is measured by various means. In general, the curves exhibit velocities that are unexpectedly greater than modeled and lead to what are flattened contours. A typical plot of tangential velocities (calculated and measured) is shown in Figure 33.

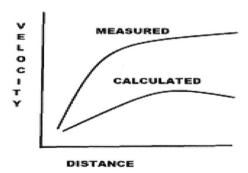

FIGURE 33

The method of analysis used *conventionally* is strictly Newtonian and one of the most widely accepted explanations is that a kind of stuff, called dark matter," exists, which doesn't show up to the usual methods of observation, that is by emitting electromagnetic energy. The presence of this extra dark matter, which only interacts gravitationally, is that it brings the measurement results back into agreement with Newtonian theory. A difficulty with this explanation is that no one has figured out what dark matter is made of.

In contrast the method of analysis used by GEQ hews to the idea of a rigid body in the extended sense but has to consider extra forces able to act on individual bodies (stars) making up the galaxy. This is true because, unlike a rigid body in the restricted sense, where the parts are "locked into place" by surrounding materials, other forces can change the dynamics of the geometry.

In particular take a look at the plots of rotation speeds depicted in Figure 33 and compare it to the rotation curve indicated in Figure 32. It may well have been that at some time in the distant past the distributions of rotation speed may have been as indicated in the lower plot of Figure 33, but, not surprisingly, over time forces have caused the rotation speeds to change in some way because the different portions of the galaxies can move relative to one another.

Several questions:

1. Is the GEQ notion that a rigid body (in the broader sense) starts out looking like the lower curve in Figure 33 and migrates to the upper curve in Figure 33 because of extra forces from some distribution of field intensity in the structure? Here I think the answer is yes.

2. If so, does it follow that one needs extra mass (dark matter) to account for that? Here I think the answer is in the negative because the final velocity distribution is dependent on a combination of the initial state of velocities and the subsequent changes from additional forces acting on the various elements of the entity. As you will see in chapter 14 it turns out the final velocity distribution is the sum of the initial linear pattern and the Newtonian pattern adding up to the upper curve in Figure 33. Details are provided in chapter 14.

Chapter 15 focuses on what we see of distant galactic motions from our view here on earth. The present view (2020), there is a monotonic increase in recession velocity as the distances increase, and most dramatically, the relatively nearby recession velocities appear to be accelerating more rapidly than distant galaxies. Why this should be happening is not clearly understood but gives rise to a mysterious extra force labeled *dark energy* or alternate names, but meaning the same thing, the cosmological constant or vacuum energy. Whatever the name, its main characteristic is that it acts in the opposite sense of gravity; it pushes instead of attracting.

What the GEQ approach takes to explain the measurements is to apply the rigid body notion to the data where the difference between equations 6 and 7 plays an exceeding important role in understanding the dynamics without the need for dark energy.

One potential conceptual problem might come from the need that angular momentum should be preserved in the model for broadly defined rigid bodies and as we look out at the distant galaxies, we see no such systemic rotations. However there are at least two reasons why we should not observe such motions. First is that the basic concept suggests the angular momentum should line up as in Figure 32; and second, the distances involved are so great that any such rotations as might arise from unaccounted forces would be so small that our intervals of observation are too brief for us to measure such motions.

One final comment that applies to chapter 15 is that one important assumption in GEQ is that space is passive, not an active participant in cosmological behavior. This is in contrast with the GR view where Einstein's development explicitly includes spatial behavior in the dance. Indeed his very first assumption is that mass causes space to curve.

This difference ends up by GEQ rejecting the existence of velocities exceeding the speed of light, an actuality in GR. The important consequence you will encounter in chapter 15 is the method of describing the relationships between the velocities of distant galaxies and distances from the earth. In GR their depiction is that such superluminal velocities occur while the GEQ view is that they can't.

Accordingly in GEQ a transformation from Special Relativity is used that constrains all velocities to be less than the speed of light. This turns out not to be required in chapter 14, where velocities are always very small compared to light velocity.

SUMMARY OF CHAPTER 14

The so-called spiral galaxy rotation problem arises from the observation that the distribution of circular velocities as you go out from the center to the outer rings does not follow Newton's law as expected; that is decrease with distance.

The discrepancy was first discovered in the mid 1970's by American astrophysicist Dr. Vera Rubin. Initially her claims were met with skepticism, but later measurements by her and others convinced the community that this in fact was true. The measured patterns are termed flattened, because they tend to look like the upper plot shown in Figure 33.

All of the attempted explanations from the 1970s to the present have focused on some variant of Newton's gravity law or the existence of some esoteric substance called dark matter that adds to visible matter thus bringing the lower curve in Figure 33 into agreement with the upper curve. The approach using a modified Newton's law, called MOND, says at very low mass situations, the law takes a form leading to the upper curve without the need for dark matter.

The difficulty for each of these solutions are that for dark matter, no one has been able to figure out what kind of matter exists that has gravitational properties, but does not show up by emitting electromagnetic energy, and for MOND, that it is entirely ad hoc, has no theoretical underpinnings, and was invented just to explain this phenomenon.

The GEQ solution takes a very different path, claiming that the measured curves are a combination of Newtonian forces and the behavior of angular momentum characteristic of rigid bodies in which circular velocity within the structure increases linearly from the center out to the maximum radius.

According to this view GEQ postulates that structures, such as galaxies, at some time in the distant past were all physically connected and separated into their present granular forms (discrete stars and gases) from an inherent property of the expansion process that caused them to break up, but retain their angular momentum in orderly fashion, greater tangential velocity the farther out they were in the original monolithic structure. GEQ refers to such distributed structures as "rigid in a broader sense."

In this chapter, six spiral galaxies are analyzed according to these ideas and it is seen that within certain constraints having to do with uncertainties due to unaccounted for forces, the resultant circular patterns are very much in agreement with measurement results.

CHAPTER 14

Spiral Galaxy Problem

About 60% of the known galaxies have the property of forming logarithmic spirals as they expand. A typical formation is shown in Figure 34.

SPIRAL GALAXY *deposit photo*

FIGURE 34

Often the pitch of these spirals is at or near 12° and the same form is observed in snail shells, goat horns, spider webs, and other natural growth patterns. This might lead one to believe that nature has a special affinity for spirals, and it represents some deep truth well beyond the present investigation.

Before about 1976 when American astrophysicist Dr. Vera Rubin published a journal paper about anomalous accelerations at galactic scales, particularly in the Andromeda Galaxy, it was believed that the orbital velocity distribution in spiral galaxies decreased with distance from galaxy center in accordance with Newton's inverse square gravity law. Her finding that the velocities did not follow his law but remained more or less constant with distance, initially was met with skepticism, but later work by Rubin and others showed her initial work was correct. This effect is now referred to as *flattening of spiral galaxy curves* and *the galaxy rotation problem.* A typical plot of the difference between what was earlier calculated for velocities versus what Rubin had claimed and was verified by measurement is shown in Figure 35.

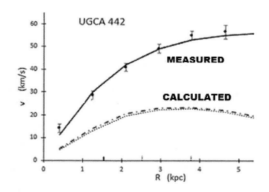

ROTATION CURVES FOR FOR UGCA 442

FIGURE 35

The plot in the upper curve is not theoretical but shows circular orbital velocities varying with distance from the galaxy center as measured. The lower curve represents what is expected based on the amount of matter and distribution of it according to observation of electromagnetic energy radiated from the galaxy.

Thus far (2020) no entirely satisfactory explanation for the difference between the upper curve and the lower one has been found. The most widely accepted one is the presence within the galaxies of *dark matter*, an invisible substance that adds to the gravitational effects of usual matter, but does not interact with electromagnetic energy, hence is "invisible." Given that it exists, its net effect is to restore the Newtonian view. If this is the true explanation, the upper curve in Figure 33 includes dark matter. However, there are problems with the dark matter hypothesis. The main difficulty is that in spite of broad efforts to determine what particles combine to make-up dark matter, no reasonable solution has been found

A second possible explanation is the MOND theory by the Israeli physicist Mordehai Milgrom who published this idea in 1983. MOND (Modified Newtonian Dynamics) posits that gravitational forces, "*...experienced by a star in the outer regions of a galaxy was proportional to the square of its centripetal acceleration (as opposed to the centripetal acceleration itself, as in Newton's second law)*." According to Milgrom's idea, this only occurs at extremely small accelerations, "*far below anything typically encountered in the Solar System or on Earth.*" This explanation has also been widely studied, but because it is totally ad hoc, no basis in theory or logical reasoning other than that it works, it is not a very satisfactory solution.

A third approach, called the wave density effect, was postulated in the mid-1960s by astronomers CC Lin and Frank Shu. According to their theory the spiral arm structures arise from different speeds of rotation between the galactic disc and the spiral elements forming the arms, the slower moving stars bunch-up in regions to form the arms much as slow moving cars form traffic jams, in this case higher density regions. In essence it is a kind of resonance phenomenon where stars inboard of the arms move more rapidly and those outboard move more slowly and enter into and exit the arms as the pressure region moves outward. Their ideas appear to have value in other contexts such as the forming of the rings of Saturn.

Each of the observed spirals are approximated to be circular orbits around an observed center and the enclosed mass estimated by evidence from visual electromagnetic emanated from the enclosed region.

The estimated mass and the assumption of circular orbits in principle should enable the calculation of decreasing velocity with distance using Newton's standard law for circular

calculation of decreasing velocity with distance using Newton's standard law for circular

orbits: $V = \sqrt{\frac{GM}{R}}$. The Newtonian equation results in the calculated lower plot in Figures 35.

However, when direct measurements are made of velocity vs distance, the upper curves result. The method used to measure velocities directly is as follows.

The velocity of the materials contained in an orbit at some particular distance from the galaxy center is estimated by observing the difference between spectral shift as seen from the earth on either side of the maxima of the orbit (relative to the viewer) and one half of this shift is taken to represent the mean velocity for that particular orbit via the standard relation of Doppler Shift to velocity, including time dilation effects from SR is:

$$f(receiver) = f(source) \sqrt{\frac{1-\frac{V}{c}}{1+\frac{V}{c}}} \tag{10}$$

This data then is used to generate the upper curve in Figure 33.

GEQ SOLUTION

According to the GEQ point of view the patterns exhibited by spiral galaxies are a mix of the usual idea of Newtonian and rigid body behaviors. In as much as any solution must be guided by measurement data, not theory, the fact that observed velocities increase with distance implies the patterns are more similar to rigid body behavior as discussed in chapters 13 and 14 than to Newton's gravity law as applied in the Solar System model. Accordingly, the solution presented here follows the notion of rigid body in a broader sense than is usually used.

The most salient issue involves angular momentum wherein according to the GEQ view, every portion of galactic material inherits some degree of tangential velocity from an earlier time when it was part of a solid body (rigid body in the restricted sense) even though our present view is that all parts are separate disconnected elements making up the galaxy. The reason for this was discussed in chapter 13 where it was hypothesized that the expansion hypothesis inherently implies that large structures such as galaxies were all physically connected in the distant past. Let me quote a portion of that discussion.

All large bodies (planets, stars, galaxies) possess angular momentum. Why this is true is not known, but the usual assumption is that at the time these objects were formed, the particles that accreted in their formations possessed rotations and passed that energy to the end product. Be that as it may, for any solid, truly rigid sphere, one understands the rotation speed of any sample taken at some arbitrary radius of the sphere is simply:

$$Tangential\ speed\ V_x = R_x/R_{max}(V_{max})$$

That is there is a straight line-up of speeds at depth as indicated in Figure 32.

This is very different than equation 9 and due to very different considerations; simply that the surrounding mass carries it along.

But in the case of a rigid body in the broader sense no such physical contact exists so apparently what must be true is that at some time in the past they had been in physical contact and even after they separated each "hunk" retained its angular momentum. This gives rise to the monotonic increase in rotation speed as they get further from the reference center.

But why should they separate? And this gets me to factor (2).

Unlike Guth's notion of spatial stretching or expansion the GEQ view is that it is above all else resident in the behavior of mass, so it has a definite affect locally. One of those effects arises from equation 3: Specifically that all matter expands at nearly the same rate but there is differential velocity between elements making up the subject mass. Specifically that any two elements of the mass where one is a bit further out from the center, will have a slight differential velocity between them, the further out one going a bit faster, so gaps will develop throughout the distribution of elements making up the total body.

Given this it is not too difficult to understand that as time goes on what started out as a single connected entity subdivides into smaller entities retaining angular momentum (speed of rotation) inherited from the original whole characterized by their locations.

To go through the details of this process is well beyond my capabilities and the immediate needs of this book, so I will just assume it is a viable explanation of how planets, galaxies and what else have you, form, where velocities increase monotonically with distance.

However there is an important difference between rigid bodies in this extended sense, as compared to what we usually mean when we talk about a solid entity, in that elements (individual stars) of the galaxy have extra degrees of freedom because they are not locked into place by adjacent portions of the structure. Because of this, their tangential (circular) motions can and evidently do change over time from that typical of a rotating solid sphere.

In order to analyze the dynamics of this system I need a physical model to apply the presumed evolution. If I imagine a spinning rigid disc, I understand the tangential velocity increases linearly with distance from the center according to the simple law, stated in equation 10. If I further imagine the disc takes on the property that any mass is free to displace, but no extra forces are present to cause it to do so, the distribution of velocities are unchanged so long as the spin rate remains constant.

The question now is, what happens if I introduce an extra spin by rotating the entire coordinate system so the outermost mass at R_{max} increase to a greater spin rate? This is analogous to the situation of an individual standing on a moving walkway (think airport) where velocities add. In this present situation, velocities likewise add, but the magnitude of the increase in speed depends on the distance from the center.

Given this model, I understand that if I am not in the rotating coordinate system and I measure the tangential velocity of some mass an arbitrary distance from the center, then I will find it is equal to the sum of two spins: one from the calculated spin of the small mass and the other from the spin of the coordinate system. These theoretically add up to the measured value .In this scenario I am positing that the spinning of the coordinate system represents an extra Newtonian force mainly caused by the observed mass, thus changing the original geometry characterized by the linear arrangement of matter into progressively more rapid tangential speed as its position lies outward from the center. It has to be assumed that the new velocity distribution is not exactly the sum of the two spins due to chaotic interactions among the various elements (stars and gases) making up the galactic ensemble.

This is pictured in Figure 36 where the spin velocity of the coordinate system is represented as a straight line terminating at V_{max} at the outermost distance, and the corresponding velocity near the center. This approximates the "best fit" spin rate for the rigid disc. Table V lists the subsequent velocity values for UGCA 442, the most important being the sum values (compared to the measurement curves) that are indicated by small squares. Figures 37 through 41 and accompanying Tables supply similar data for additional spiral galaxies.

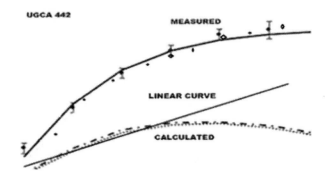

DATA PLOT FOR UGCA 442

FIGURE 36

TABLE V

MEASURED DATA FOR UGCA 442

DISTANCE (kpsec)	ROTATION SPEED (calculated)	ROTATION SPEED (linear)	ROTATION SPEED (sum)	ROTATION SPEED +/5 (measured)
1 kpsec	11 km/sec	7 km/sec	18 km/sec	22 km/sec
1.5 kpsec	15 km/sec	15 km/sec	30 km/sec	33 km/sec
2 kpsec	20 km/sec	18 km/sec	38 km/sec	40 km/sec
2.5 kpsec	21 km/sec	20 km/sec	41 km/sec	45 km/sec
3 kpsec	23 km/sec	23 km/sec	46 km/sec	48 km/sec
3.5 kpsec	23 km/sec	24 km/sec	47 km/sec	51 km/sec
4 kpsec	23 km/sec	28 km/sec	51 km/sec	54 km/sec
4.5 kpsec	22 km/sec	32 km/sec	54 km/sec	55 km/sec
5 kpsec	22 km/sec	36 km/sec	57 km/sec	57 km/sec

Now I repeat this data for additional galaxies starting with UGC 128.

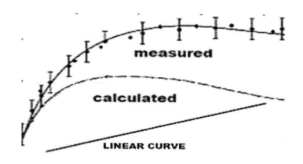

DATA PLOT FOR UGC 128

FIGURE 37

TABLE VI

MEASURED DATA FOR UGC 128

DISTANCE (kpsec)	ROTATION SPEED (calculated)	ROTATION SPEED (linear)	ROTATION SPEED (sum)	ROTATION SPEED +/10 (measured)
5 kpsec	60 km/sec	14 km/sec	74 km/sec	71 km/sec
10 kpsec	80 km/sec	20 km/sec	100 km/sec	100 km/sec
15 kpsec	85 km/sec	28 km/sec	113 km/sec	120 km/sec
20 kpsec	85 km/sec	33 km/sec	118 km/sec	130 km/sec
25 kpsec	85 km/sec	40 km/sec	125 km/sec	130 km/sec
30 kpsec	80 km/sec	45 km/sec	125 km/sec	130 km/sec
35 kpsec	70 km/sec	55 km/sec	125 km/sec	130 km/sec
40 kpsec	65 km/sec	60 km/sec	123 km/sec	130km/sec

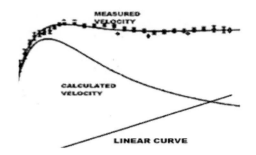

DATA PLOT FOR NGC 3198

FIGURE 38

TABLE VII

MEASURED DATA FOR NGC 3198

DISTANCE (kpsec)	ROTATION SPEED (calculated)	ROTATION SPEED (linear)	ROTATION SPEED (sum)	ROTATION SPEED +/5 (measured)
5 kpsec	135 km/sec	14 km/sec	149 km/sec	145 km/sec
10 kpsec	122 km/sec	30 km/sec	152 km/sec	152 km/sec
15 kpsec	100 km/sec	42 km/sec	142 km/sec	145 km/sec
20 kpsec	86 km/sec	53 km/sec	139 km/sec	145 km/sec
25 kpsec	75 km/sec	65 km/sec	140 km/sec	145 km/sec
30 kpsec	67 km/sec	78 km/sec	145 km/sec	145 km/sec

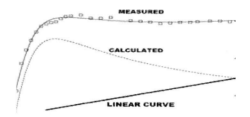

DATA PLOT FOR NGC 4728

FIGURE 39

TABLE VIII

MEASURED DATA FOR NGC 4728

DISTANCE (kpsec)	ROTATION SPEED (calculated)	ROTATION SPEED (linear)	ROTATION SPEED (sum)	ROTATION SPEED +/5 (measured)
5 kpsec	125 km/sec	25 km/sec	150 km/sec	152 km/sec
10 kpsec	110 km/sec	40 km/sec	150 km/sec	150 km/sec
15 kpsec	100 km/sec	50 km/sec	147 km/sec	150 km/sec
20 kpsec	90 km/sec	55 km/sec	145 km/sec	149 km/sec
25 kpsec	78 km/sec	68 km/sec	146 km/sec	150 km/sec
30 kpsec	75 km/sec	75 km/sec	150 km/sec	150 km/sec

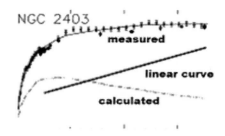

DATA PLOT FOR NGC 2403

FIGURE 40

TABLE IX

MEASURED DATA FOR NGC 2403

DISTANCE (kpsec)	ROTATION SPEED (calculated)	ROTATION SPEED (linear)	ROTATION SPEED (sum)	ROTATION SPEED +/5 (measured)
2.5 kpsec	60 km/sec	40 km/sec	100 km/sec	100 km/sec
5 kpsec	60 km/sec	50 km/sec	110 km/sec	115 km/sec
7.5 kpsec	63 km/sec	60 km/sec	123 km/sec	125 km/sec
10 kpsec	52 km/sec	70 km/sec	122 km/sec	129 km/sec
12.5 kpsec	49 km/sec	80 km/sec	129 km/sec	130 km/sec
15 kpsec	40 km/sec	90 km/sec	130 km/sec	130 km/sec
17.5 kpsec	35 km/sec	100 km/sec	135 km/sec	130 km/sec

152

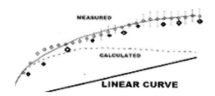

DATA PLOT FOR NGC 247

FIGURE 41

TABLE X DATA FOR NGC 247

TABLE X DATA FOR NGC 247				
DISTANCE (kpsec)	ROTATION SPEED (calculated)	ROTATION SPEED (linear)	ROTATION SPEED (sum)	ROTATION SPEED +/5 (measured)
1 kpsec	45 km/sec	0 km/sec	45 km/sec	45 km/sec
2 kpsec	53 km/sec	2 km/sec	57 km/sec	60 km/sec
3 kpsec	58 km/sec	8 km/sec	68 km/sec	70 km/sec
4 kpsec	62 km/sec	14 km/sec	76 km/sec	80 km/sec
5 kpsec	61 km/sec	20 km/sec	81 km/sec	85 km/sec
6 kpsec	65 km/sec	26 km/sec	91 km/sec	92 km/sec
7 kpsec	64 km/sec	32 km/sec	96 km/sec	95 km/sec
8 kpsec	63 km/sec	38 km/sec	101 km/sec	99 km/sec
9 kpsec	62 km/sec	44 km/sec	104 km/sec	100 km/sec
10 kpsec	60 km/sec	50 km/sec	110 km/sec	110 km/sec

Finally let me make clear that while this analysis from the GEQ point of view is pleasing in that it neither requires the invention of dark matter, nor does it require an ad-hoc change to Newton's gravity law with no theoretical basis, it does require the assumption about the evolution of galaxies from solid entities into their distributed discrete forms, and certainly this notion requires a careful evaluation as to its merits and possibility. Even more, as discussed in the essay on truth in physics appearing at the end of this book, the over-riding requirements for validity go well beyond any of the materials presented in this chapter.

SUMMARY OF CHAPTER 15

One of the big topics investigated by cosmologists has been and continues to be our physical relation to the rest of the universe. Einstein's early work in 1905 on Special Relativity provided us with well-defined measurement tools for addressing the issue, and his later work in 1915 on gravity (General Relativity) gave us a theoretical model for framing the quest that has been very fruitful, although not always along the lines Einstein originally thought they would go.

In 1912, Vesto Slipher measured red shifts in light from distant galaxies and claimed this meant those galaxies (at the time called star islands) were all moving away from us. As sometimes happens in the sciences as well as in other human endeavors, his announcement was greeted with a big yawn, "So what," no one in the community cared.

In 1915, Einstein published his paper on gravity (GR) that in its original form said the universe should be expanding, but he apparently didn't know about Slipher's work, so he added in a constant such that his theory said the universe was constant in size.

Then two more discoveries were made. In 1922, a set of equations by a theoretical physicist, Alexander Friedmann, that implied the universe might be expanding, and in 1929 measurements by Edwin Hubble, an American lawyer/cum astronomer, who announced that measurements showed all galaxies were moving away from us—in every direction—and the further away they were, the faster they were going.

Needless to say, Einstein removed that constant he had added (he later called it his greatest blunder), and everyone in the community agreed Slipher had been right after all. *But, dear reader, Einstein's constant, called the cosmological constant, reappears, so don't forget it.*

The method used by astronomers to measure distances are complex but involve the way various kinds of stars explode at the end of their lives and I will not go into details here. If you want to know, read chapter 15 or Google it.

The next big issue was, and still is, to find a value for what is called the Hubble Constant, a measure of recession velocities as a function of distance.

Currently there are three ways of measuring it; two agree, the third disagrees being somewhat smaller but smaller enough to cause problems. All three numbers are expressed as km/sec per Megaparsec. The present "best value" is about 74 units, the nasty third one is about 67 units.

The most definitive work on velocity versus distance was carried out by a group of physicists headed by Saul Perlmutter, Brian P. Schmidt, and Adam Riess in 1999. They were awarded the 2011 Nobel Award for the results. The big discovery was not only the universe expanding, but its present expansion rate is accelerating compared to what it was in the past. This gives rise to the idea of dark energy, the same thing as Einstein's cosmological constant in the sense it operates the reverse of gravity: It repels instead of attracting. So, here is his constant again, back to life.

What is meant by an accelerated expansion is that the distances between galaxies gets greater in time and space. If you think of a plum pudding while it is rising and note the plums are getting further apart as the pudding gets larger, that's "expansion." But if you also notice that the distances get larger for plums closer to the center as the expansion goes on than those that moved apart earlier, than you can say the expansion is accelerating.

Finally if you have a larger enough pudding so you can see earlier spacings (way out from the center), ones a medium distance out, and latest ones near the center, and the spacings for the closest in ones are further apart than expected than any of the other ones, you can say the acceleration is increasing in space and time. Whatever is causing the acceleration is increasing in magnitude.

Another simpler way of envisioning this last stage is to imagine two race cars side by side accelerating together down the road. If one of the drivers presses harder on his accelerator, he will accelerate at a greater rate and pass the other slower car, but both are still accelerating.

All this above brings you up to date on the conventional view of the cosmological dynamics.

Now let me look at what GEQ has to say about all of this, focusing on the goal of getting rid of dark energy by showing the accelerated expansion is actually due to properties of metrics ,as suggested by GEQ considerations.

There are several steps needed to accomplish this and success is accomplished when and if the GEQ approach obtains a result in agreement with the Riess et al measurements without dark energy. There are two reasons success is defined this way: First, dark energy is an ad hoc invention to explain what is measured and physics does not like ad hoc inventions; and second, measurement is the gold standard for truth so one tends to take measurement results at face value.

The first step is to deal with the apparent increase in acceleration as time passes. In standard cosmology a linear scale is used such that, taking the race cars as an example, the increased acceleration of the faster car shows up as an extra linear increase in its velocity relative to the slower car. In GEQ a more proper nonlinear scale from Special Relativity (SR) is used that comes from the idea that nothing can reach the speed of light. Since the recession velocities of galaxies are so large, the net effect is that most or all of the extra velocity vanishes. *In the GEQ analysis done here, Riess's velocity data is translated into the SR values for further analysis/comparison with GEQ methods.*

The second step is to correct Riess distances to consider differences in distances as seen in one coordinate system when looking at distances in a different coordinate system. According to the methods developed in SR, when an observer looks at distances as specified in a different coordinate system, he has to use a certain equation (not the same one as in step 1) to get proper values. In this particular situation, where the views are shifted in time ("now" looking at "past"), they are different coordinate systems. *In the GEQ analysis done here, the Riess distance data is translated into SR values for further analysis/comparison by GEQ methods.*

)The third and last step is to calculate the GEQ values for velocity and show that they agree with the Riess data from step one. This is accomplished as described below.:

Noting that recession velocity increases monotonically with distance, GEQ points out that this collection of galaxies behaves like a rigid body in the extended sense, just as was true for spiral galaxies in the previous chapter. For any such rigid body, we see no change in size over time except in cases where sizes and times are great enough that the difference between equations 6 and 7 come into play.

In situations like that, we do see the so-called rigid body increase in size, because the *ratio of sizes independent of density* is no longer true and accordingly, we see the sizes change.

In this particular case we can find the magnitude of that change by expressing the difference between equations 6 and 7 as an acceleration and multiplying it by the length of time for each of the distance from the observer on earth to the galaxy he is observing. That gives him the recession velocity for each galaxy. He gets those time intervals by dividing the translated distance in step two by the speed of light.

When all this is done, the comparison of velocities from step one and step three show they are identical and the GEQ method does not use dark energy to get the answer. Q.E.D.

KEY CONCEPTS FOR CHAPTER 15

1. **THE UNIVERSE IS HOMOGENOUS AND ISOTROPIC**- Every observer sees uniform density on large scales (homogenous) in every direction (isotropic). This is true no matter where the observer is. What is implied by this is that there is no center to the universe. A simple two-dimensional model is the surface of an expanding balloon. Note that the balloon doesn't have to be spherical for this analogy to hold. A simple three-dimensional model is a plum cake baking wherein the plums move apart as the batter grows in size because of "rising."

2. **THE UNIVERSE'S SIZE IS INFLATING**- Inflation is an idea first proposed by Alan Guth and others around 1980. According to that theory, immediately following the Big Bang, the size of the universe rapidly increased for a very short time. Since then, it has continued to increase at a much slower rate. He proposed the energy for this expansion comes from vacuum energy, which is the same as dark energy. According to the inflation theory, close together groupings below the galactic scale do not participate in this expansion but remain as units carried outward in the spatial expansion.

3. **RED SHIFT**- Many elements such as hydrogen have well known spectra (characteristic distributions of lines in displays of their frequencies) that can be observed at certain locations in visual displays. When their shifts are to lower frequencies, they are said to be red shifted, or to higher frequencies blue shifted. In the case of red

shifted it means the light source is moving away from the observer, the amount of shift is due to the velocity difference between the observer and the source. The same thing is true for blue shift except then the source is moving towards the observer.

CHAPTER 15

Hubble Constant, And Geq

In 1912, Vesto Slipher measured red shifts in light from distant galaxies and claimed these shifts were due to Doppler effects from the galaxy's motions away from us. Neither he nor any other astronomers of the day appreciated the importance of this observation.

In 1915, Einstein published his celebrated new theory of gravity (GR) that in its original form required the universe be continuously expanding. But he and the rest of the physics community, either ignored, had forgotten, or didn't believe Slipher's claim, so he inserted a factor, called the cosmological constant, into his equations, which exactly cancelled out the expansion. So in his view the universe was static in size.

Then along came the Friedmann equations (Alexander Friedmann) in 1922, based on the observation that the universe appeared homogeneous and isotropic. His results implied the universe might be expanding and that Einstein's field equations in conjunction with his analysis enabled a means to calculate its magnitude. Again, in 1927, Georges Lemaitre independently developed similar solutions to Einstein's equations.

Finally, in 1929, the American astronomer Edwin Hubble, announced his discovery that galaxies from all directions, appeared to be moving away from us and that the motions, just as Slipher had earlier claimed, were indicated by red shifts. This resulted in Einstein abandoning his cosmological constant saying that this was "my biggest blunder." (*Still later, a book by Donald Goldsmith titled, Einstein's Greatest Blunder? discusses the not-so-simple consequences of that cosmological constant, which has since come back to life.*)

What Hubble observed was that the brightness of the star/objects seemed to be closely correlated with the degree of red shift, dimmer objects having greater shifts. This suggested to him that for star clusters, well beyond the size of our own galaxy, *most* were moving away from us in orderly manner. That implied one could represent something about universal expansion as indicated in Figure 42. At the time, the grouping of stars was generally called star islands, but now we label them as galaxies.

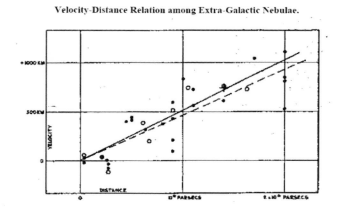

FIGURE 42

While the relationship between velocity and red shift was well understood by 1929, the labeling of the " y" axis with velocity presented no problem, but the actual distances indicated on the x axis were (and still are) a more complicated issue. The method of scale calibration available to him came from work Henrietta Swan Leavitt of Harvard College (Cambridge, MA) had done in studying the dynamic brightness of what are called Cepheid variable stars. She discovered that they obeyed a period-luminosity relationship that was longer for intrinsically more luminous stars (1900-1913).

This gave Hubble a means to measure distance by measuring periodicity of many Cepheid stars, determine their actual luminosity and from their apparent brightness m their distances. In principle, this enabled him to apply a distance scale to the "x" axis.

Using these methods from 1923 through 1929, Hubble and others mainly used the 100-inch Hooker Telescope on Mt. Wilson in California, to develop the data plotted in Figure 42. From this, he concluded:

1. The galaxies all were moving away from us at velocity's thatincreased with distance,
2. On average one could assign a value of 500 km/sec per Megaparsec distance where a Megaparsec = 3×10^{22} meters,
3. That a law (now called The Hubble Law) is

$$V = H_o d.$$

Where V is the velocity of recession, d is the distance, and H_o is the Hubble constant. The subscript "o" is added to "H" indicating that its value can change over time intervals measured in eons, but this is its present value.

Establishing value of H_o The work accomplished by Hubble, Slipher, Leavitt, and others from roughly 1900 through 1930 convinced the astronomy/ cosmology communities that the Universe was expanding, that Einstein's use of his cosmological constant was in error (at least how he used in GR), but it also raised many questions about how to explain this new view.

For one thing if you take the reciprocal of the Hubble constant,

$(1/H_0)$(km/Mpsec*seconds) you get the age of the universe to be: 1/500 km/Mpsec*seconds = 2 billion years.

The problem is that we know of stars in the universe on the order of 13 billion years, so it follows that the original value for H$_o$ was way too high. Why did Hubble get it so wrong? The reasons are complex, and later measurements, made in 1956, dropped the value to about 180, (Humesan, Mayall and Sandage) , then in 1958 to about 75 (Allan Sandage) and so on as the true value became more and more into question.

Here is the situation currently.

There are three methods used to calibrate distances from the earth to very far away stellar objects.: The first is by using what are called, *standard candles* , the second method involves using the *lensing effect*, the third using *the cosmic microwave background* that formed just after the Big Bang birth of the cosmos, which is now estimated to have occurred 13.8 billion years ago.

In the use of so-called *standard candles* the method is similar to that used by Hubble, except the standard candles are often provided by type Ia supernovae, which are white dwarf stars that have reached their stability limits and explode. (Mass greater than about 2.765×1030 kg, also known as the Chandrasekhar limit.) When this happens the time-profile of the explosion is closely related to its maximum luminosity, thus providing a clue as to its distance from the earth via its apparent brightness. While the period from the time of an explosion through its brightness decay is relatively short (on the order of months), enough of them occur and are detected as to make their very well-known brightness curves. In addition to supernovae, variable Cepheids have once again come into use with better understandings of their behavior now than earlier for Hubble and also the behavior of giant red stars can be and are used. The latest value obtained using this method (2019) is **74 km/sec-mpsec.**

A recent new method (2019) used lensing effect for far-away quasars and arrived at a result of **73 km/sec-mpsec,** in essential agreement with the standard candle method.

Finally with a method using known or assumed properties of *cosmic microwave background radiation,* cosmologists have been able to develop expansion models and from this predict the value of H_o. The best value measured this way is **67.4 km/sec-mpsec**. Current understanding of the of cosmic microwave background radiation is such that all the different methods should give much the same results. Why they don't is a topic of on-going interest.

In 1999, studies by Adam G. Riess used Type Ia supernovae to show that not only is the universe is expanding, but the rate of expansion is currently accelerating. This was entirely unexpected and Riess and two other members of the research group have received Nobel Prizes for their discovery. Let me take a look at their result, which is subtle and plays an important role in how GEQ looks at these materials, especially the Hubble constant.

First of all, keep in mind that the distances to far-away galaxies are very great, so any data collected about them by using telescopes is from the past. That is the distances and velocities we measure are from *where they were a long time ago*. So if I say for example, "galaxy X is accelerating away from us", it means the velocity I see galaxy X moving at (what I see now) is different than what I would see if I magically went back in time and see it's velocity when it was closer to us.

In this regard, note that the definition of H is (velocity-change/distance) so the plot of distance and time versus velocity shown in Figure 43 below has distance along the x axis increases to the right, but later times ("our time") to the left.

Before Riess and colleagues made their measurements, the expectation was that they would see more distant galaxies decelerating because gravity should have slowed the expansion over time. That is not what they measured. they found just the opposite appeared to be true. Let me represent that in terms of the Hubble constant/flow as in Figures 43a and 43b.

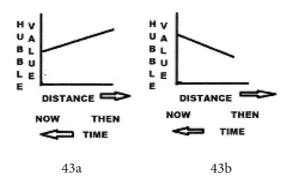

43a   43b

FIGURE 43

This is very exaggerated to make clear what he observed. Figure 43(a) shows the Hubble constant greater to the far right (the past), decreasing in value as time came closer to "now," and this is what he expected. Figure 43(b) indicates what he saw, just the opposite, smaller early on, then greater now.

Enter *dark energy* and Allan Guth's notion of *"inflation"* mentioned in the key concepts section at the beginning of this chapter. Riess and colleagues' findings suggest the reason for these results is that space itself is expanding more rapidly now than it was in the past and carrying the galaxies along with it. The energy for this process is thought to be dark energy or the same thing with several different names (Einstein's cosmological constant with new clothes, zero-point energy, or vacuum energy). Call it what you like, but it has the property of acting in the opposite direction of Newton's gravity force. The big open question is, what is causing this greater acceleration now, relative to what it was in the past.

Before I move on to how GEQ interprets these findings, let me take a quick look at the details of Riess's 1999 measurements and show you a plot of his results.

There are two pieces of information available from long distance observations of distant galaxies and standard candles, the method used by Riess. The apparent brightness of Type Ia supernova in the immediate vicinity of some subject galaxy provides a distance measure and the red shift of light from the receding galaxies provides its velocity.

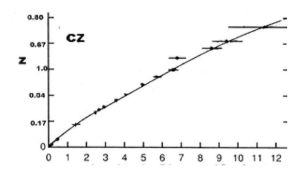

RIESS DATA PLOT (from Ned Wright data plot)

FIGURE 44

Notice several things about this plot.

1. The y axis is labeled cz km/sec where z goes roughly from 0.1 to 6.0 making the plotted velocity values vary from about 1 x 105 km/sec to about 5 x 105 km/sec. But according to special relativity nothing can go faster than the speed of light (3 x 105 km/sec). What's going on here?

What is going on is that the "framing" of the plot is in terms of Guth's inflation hypothesis and there is nothing in SR or GR that says "space" expansion velocity cannot exceed the speed of light.

Taken with that understanding, the plots of far-away galaxies can exceed the velocity of light from our point of view. If we were adjacent to each of those galaxies plotted here, we would discover their "local velocities" relative to other nearby galaxies would be less than c. The z as used in this graph refers to the *observed* spatial velocity relative to the speed of light. When z = 1, the recession velocity is equal to the speed of light. In effect it is a plot of z.

Note that the earlier part of plotted data is more or less a straight line indicating the spatial expansion is accelerating at a constant rate, while the later plot (starting around 1 Gpc) starts to curve a bit downward, indicating that while the expansion is still accelerating, the acceleration is decreasing somewhat with distance. This of course is what Figure 25 shows in exaggerated detail.

GEQ SOLUTION

The main solution suggested by GEQ centers on the same concept presented in chapters 13 and 14, that the galactic clusters taken as a group behave as though they are a rigid body in the broad sense. This view as you will see is very important.

Before I start the analysis, let me interject an important underlying issue: that the GEQ explanation for the birth of the universe starting with something like the Big Bang requires some discussion.

The notion of the Big Bang being similar to an explosion is specifically excluded in the standard cosmological view. There are several reasons for this rejection. Collectively the clues/requirement came from the Freidman equations, Slipher/Hubble observations of red shift, the belief in GR (once Einstein removed his cosmological constant), and a need for any theory to include the experimental observation that the universe appears homogeneous and isotropic. The inherent nature of an explosion is that it occurs at a point in space, thus conflicting with the twin requirements of homogeneity and isotropy. Guth's notion of "inflation of space" took care of that difficulty. This notion of homogeneity and isotropy at a large enough scale is called the cosmological principle.

It remains a requirement of the GEQ notion and in view of its rejection of the idea that space is anything more than a passive background in the dance matter does, it does not allow for Guth's explanation. Some other means must be found that allows the cosmological principle to remain whole.

The two approaches I suggest do start with something very much like an explosion, but at least in the first model, it is not an explosion in the usual sense that consists of the rapid expansion of some form or forms of material objects be they gases or minuscule particles of some sort, but rather several stages following:

First, a process in which some portion of the existing matter, not having reached escape velocity relative to one another, coalesces by said expansion into a single entity.

Second, continuous expansion causing contained matter to heat to an extent that all contained matter dissociates its molecular bonds and becomes some-sort of "pre-matter".

Third, the energy density becomes so high as to cause a rapid expansion into empty space (no work done) so the resulting "miasma" is more-or- less of uniform energy density and it is unclear whether such a process can be said to have a center.

Fourth, a degree of non-uniformity of some portions of the miasma so a high probability of reforming molecules locally from the miasma, and

Fifth, a re-establishment of a "local" universe that for all possible measurements obeys the cosmological principles but has less total energy than the parent universe it arose from.

The second proposed model starts somewhat later in the evolutionary process where matter as we know it already exists and complexes of congregated materials (stars) actually explode. If many such events occur over time, the locality of any one event is destroyed and what results is a more or less homogeneous and isotropic distribution of matter.

Are either of these fantasies possible? Of course, I do not really know. What I do know is that if GEQ is correct, that space is passive, then something happened that led to matter distribution appearing homogeneous and isotropic without Guth's notion of *inflation*. In the following analysis I take that as a given. As for the differences between the conventional and the GEQ views:

First, the plot of velocity is constrained by considerations from Special Relativity where nothing can reach or exceed the velocity of light, so the plot must be in the form of Figure 45 and the notion of superluminal velocity vanishes.

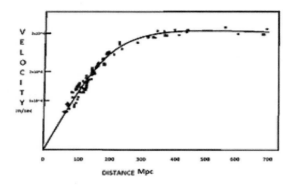

PLOT OF VELOCITY VS LARGE COSMO-
LOGICAL DISTANCE FOR GEQ

FIGURE 45

Accordingly I have to translate the Riess data into the values they would get if they used the SR equation corresponding to Figure 45. This is necessary because this is the format used by GEQ. Their data, shown in Figure 45 plotted in terms of z, is listed in Table XI, where I have added additional data below 700 mpsec from a source set measured at the same time as Riess's data for shorter distances. The SR equation used for this data set conversion is:

$$\sqrt{\frac{1+\frac{v}{c}}{1-\frac{v}{c}}} \; -1 = z \tag{14}$$

TABLE XI VELOCITY VS. DISTANCE RIESS DATA

DISTANCE and Z		VELOCITY km/sec
50 mpsec	.01	2.98×10^3
100 mpsec	.02	5.94×10^3
200 mpsec	.04	1.18×10^4
250 mpsec	.05	1.46×10^4
300 mpsec	.06	1.74×10^4
400 mpsec	.08	2.31×10^4
500 mpsec	.10	2.85×10^4
600 mpsec	.12	3.38×10^4
650 mpsec	.13	3.65×10^4
1.0 gpsec	.17	4.63×10^4
2.0 gpsec	.33	8.32×10^4
3.0 gpsec	.50	1.15×10^5
4.0 gpsec	.67	1.42×10^5
5.0 gpsec	.80	1.58×10^5
6.0 gpsec	.90	1.7×10^5
7.0 gpsec	1.10	1.89×10^5
8. 8.0 gpsec	1.20	1.93×10^5
9.0 gpsec	1.30	2.04×10^5
10 gpsec	1.50	2.13×10^5

Understand <u>this is the Riess data</u> restated in a different format. This is important because it is these velocities I will use as a comparison to the

GEQ method of calculation. The next step is to begin the GEQ calulation by adjusting for epoch. This is done by recognizing that any coordinate system viewed at a different time can be considered a different coordinate system and distances between them should be transformed by equation (15), also from special relativity. The new distances that will be used by GEQ in its calculations are listed in Table XII. In the calculation, values for velocity are taken from Table XI and time is old distance divided by c.

$$dn := \frac{1}{\left(1-\frac{v^2}{c^2}\right)^{\frac{1}{2}}} \cdot (do - vt)$$

(15)

TABLE XII

OLD DISTANCES VS. TRANSFORMED DISTANCES

OLD DISTANCE (do)	ANSFORMED DISTANCE (dn)
50 mpsec	$1.53*10^{24}$ m
100 mpsec	$3.02*10^{24}$ m
200 mpsec	$5.93*10^{24}$ m
250 mpsec	$7.343*10^{24}$ m
300 mpsec	$8.73*10^{24}$ m
400 mpsec	$1.14*10^{25}$ m
500 mpsec	$1.40*10^{25}$ m
600 mpsec	$1.65*10^{25}$ m
650 mpsec	$1.78*10^{25}$ m
1 gpsec	$2.63*10^{25}$ m
2 gpsec	$4.64*10^{25}$ m
3 gpsec	$6.18*10^{25}$ m
4 gpsec	$7.38*10^{25}$ m
5 gpsec	$8.59*10^{25}$ m
6 gpsec	$9.73*10^{25}$ m
7 gpsec	$1.03*10^{26}$ m
8 gpsec	$1.10*10^{26}$ m
9 gpsec	$1.21*10^{26}$ m
10 gpsec	$1.24*10^{26}$ m

It is important to do this adjustment in distances because the final calculation for GEQ velocities involves these modified distances and the difference between equations 6 and 7. Equation 15 incorporates what is called the "Lorentz factor" that Einstein borrowed from the Dutch physicist Hendrik Lorentz and used extensively in his 1905 paper on SR.

The next and final difference between the standard view and the GEQ view is much more subtle and involves equations 6 and 7. In summary it is a consequence of the GEQ observation that the universe behaves as though it was a rigid body in the broad sense. Let me start by considering the behavior of rigid bodies in the GEQ world.

According to the underlying principle in GEQ all matter is increasing in size in accordance with equation 6, trying to get up to light velocity "c" but never quite getting there. On the other hand, light is travelling exactly at "c" so it is in accordance with equation 7. Consequently, when I specify the length of some physical object by "*number of wave lengths*", there is a very slight difference between that measurement and a measurement using equation 6.

Which way does the difference go; is the equation 6 measurement shorter, or or the equation 7 measurement shorter?

That depends on how I define the independent variable "t" in equation 6. If I take it to be of the form I have been using throughout this book:

$$t_{eq} = \frac{R}{c}(1 - \frac{R_{miin}}{R})^{1/2} + \left(\frac{R_{miin}}{c}\right) ln\left[\left(\frac{R}{R_{min}}\right)^{\frac{1}{2}} + \left(\frac{R-R_{min}}{R_{min}}\right)^{\frac{1}{2}}\right] \quad (6)$$

$$t = \frac{R}{c} \quad \text{or} \quad R = Ct \quad (7)$$

Then I'm using t=R/c on the right-hand side of the equation as the independent variable and this implies that teq must be greater than t. If you use equation 6 to crunch a few numbers, you will quickly find this is correct.

What all of this means in cosmological measurements is that since I assume the universe behaves like a rigid body and teq > t, it says the entire scale of sizes perceived by an observer in the earth/Solar System increases in magnitude with time, but since he sees his local sizes as unchanged, he sees the universe increasing in size!

What is needed now is a set of relationships to quantify the increases seen by the observer. If you recall the method used in chapter 11 to calculate the expansion of the earth, there is every reason to assume the same technique should be used here with the difference that the focus is on distant bodies, not on the earth. In that context, I used the phrase *ruler length* to contrast with *wavelength length,* pointing out that the latter always underestimates the former and fails to capture its time- dependent nature. This is identical to the discussion contrasting t_{eq} and t, so let me follow the earlier development.

I developed a quantity I labeled distortion velocity. This is given by:

$$\frac{GM_e}{R_1{}^2}\left(t_{eq1} - t_1\right) - \frac{GM_e}{R_2{}^2}\left(t_{eq2} - t_2\right) = distortion velocity$$

Where subscript 1 refers to the observer's location on the earth, subscript 2 refers to the locations of distant galaxies, R_e is the radius of the earth, and M_e is the mass of the earth. Notice that the second term in this application goes to zero (see equation 6), so this becomes simply:

$$\frac{GM_e}{R_1{}^2}\left(t_{eq1} - t_1\right) = distortion velocity$$

In this analysis, since I am interested in the net effect on the observer in his coordinate system that includes the Sun, there are two such equations, one for the earth as shown here, and another for the Sun, so the net magnitude is the vector sum of the first two terms because the 2 terms go to zero, leaving:

$$\frac{GM_e}{Re^2}\left(t_{eqe} - t_e\right) + \frac{GMs}{Rs^2}\left(t_{eqs} - t_s\right) = distortion velocity$$

In the next calculations I intend to carry out, it is more convenient to express this term in the form for acceleration, and it becomes:

$$\frac{GM_e}{Re^2}\left(\frac{t_{eqe}}{t_e} - 1\right) + \frac{GMs}{Rs^2}\left(\frac{t_{eqs}}{t_s} - 1\right) = distortion acceleration$$

Now occurs something odd. The way I intend to calculate velocities of distant galaxies is to multiply this acceleration quantity by the time interval for each of the galaxies. Since all of these galaxies are in different angles relative to the Solar System and measurements are made at different times, one would expect no "best value" for the acceleration term since it is derived from the vector sum of quantities that surely must change in time. The experimental results used by the GEQ method shows that a "best value" of 5.86×10^{-13} km/sec^2 in fact occurs. I cannot explain why this is the case.

Accepting this result, I can obtain the predicted velocities for distant galaxies as listed in Table XIII. A sample calculation is provided to make the method used clear. Note that the times for each galaxy uses the transformed distance from Table XII (dn) to obtain the time intervals.

Example: (1.53×10^{24})/c x $(5.86 \times 10^{-13}$ km/sec^2) = 3 x 10^3 km/sec

1⁵⁹

GEQ DISTANCES AND VELOCITIES

DISTANCE (dn)	VELOCITY (GEQ)
$1.53*10^{24}$ m	$3*10^3$ km/sec
$3.02*10^{24}$ m	$5.9*10^3$ km/sec
$5.93*10^{24}$ m	$1.2*10^4$ km/sec
$7.343*10^{24}$ m	$1.4*10^4$ km/sec
$8.73*10^{24}$ m	$1.7*10^4$ km/sec
$1.14*10^{25}$ m	$2.2*10^4$ km/sec
$1.40*10^{25}$ m	$2.7*10^4$ km/sec
$1.65*10^{25}$ m	$3.3*10^4$ km/sec
$1.78*10^{25}$ m	$3.5*10^4$ km/sec
$2.63*10^{25}$ m	$5*10^4$ km/sec
$4.64*10^{25}$ m	$9*10^4$ km/sec
$6.18*10^{25}$ m	$1.2*10^5$ km/sec
$7.38*10^{25}$ m	$1.4*10^5$ km/sec
$8.59*10^{25}$ m	$1.7*10^5$ km/sec
$9.73*10^{25}$ m	$1.9*10^5$ km/sec

Now let me compare velocities I obtain this way to the Riess data in Table IX. Note have not invoked the idea of dark energy to obtain these results, but that the observed ities are very much in accordance with those to be expected for a rigid body.

TABLE XIX

COMPARISON OF RIESS DATA TO GEQ DERIVATION

DISTANCE (d0)	DISTANCE (dn)	VELOCITY (direct calculation.)	VELOCITY (GEQ)	H km/sec/ mpsec
50 mpsec	$1.53*10^{24}$ m	$3*10^3$ km/sec	$3*10^3$ km/sec	59.2
100 mpsec	$3.02*10^{24}$ m	$6*10^3$ km/sec	$5.9*10^3$ km/sec	58.9
200 mpsec	$5.93*10^{24}$ m	$1.18*10^4$ km/sec	$1.2*10^4$ km/sec	56.9
250 mpsec	$7.343*10^{24}$ m	$1.46*10^4$ km/sec	$1.4*10^4$ km/sec	55.4
300 mpsec	$8.73*10^{24}$ m	$1.74*10^4$ km/sec	$1.7*10^4$ km/sec	55.1
400 mpsec	$1.14*10^{25}$ m	$2.3*10^4$ km/sec	$2.2*10^4$ km/sec	53.1
500 mpsec	$1.40*10^{25}$ m	$2.8*10^4$ km/sec	$2.7*10^4$ km/sec	50.8
600 mpsec	$1.65*10^{25}$ m	$3.4*10^4$ km/sec	$3.3*10^4$ km/sec	50.5
650 mpsec	$1.78*10^{25}$ m	$3.65*10^4$ km/sec	$3.5*10^4$ km/sec	49.9
1 gpsec	$2.63*10^{25}$ m	$4.7*10^4$ km/sec	$5*10^4$ km/sec	40.0
2 gpsec	$4.64*10^{25}$ m	$8.3*10^4$ km/sec	$9*10^4$ km/sec	31.0
3 gpsec	$6.18*10^{25}$ m	$1.2*10^5$ km/sec	$1.2*10^5$ km/sec	26.7
4 gpsec	$7.38*10^{25}$ m	$1.4*10^5$ km/sec	$1.4*10^5$ km/sec	20.9
5 gpsec	$8.59*10^{25}$ m	$1.58*10^5$ km/sec	$1.7*10^5$ km/sec	17.6
6 gpsec	$9.73*10^{25}$ m	$1.7*10^5$ km/sec	$1.9*10^5$ km/sec	15.0
7 gpsec	$1.03*10^{26}$ m	$1.89*10^5$ km/sec	$2*10^5$ km/sec	13.0
8 gpsec	$1.10*10^{26}$ m	$2*10^5$ km/sec	$2.2*10^5$ km/sec	11.1
9 gpsec	$1.21*10^{26}$ m	$2*10^5$ km/sec	$2.3*10^5$ km/sec	9.6
10 gpsec	$1.24*10^{26}$ m	$2.2*10^5$ km/sec	$2.3*10^5$ km/sec	8.8

The last column provides the values for the Hubble constant, time- transformed to bring values in concert with the difference between equations 6 and 7. The implication is, as stated previously, that what we see as spatial expansion at the galactic level is equivalent to the expansion of a so-called rigid body.

A final observation is that on the surface this view looks very different from Guth's idea of inflation, in fact except for the claim that the expansion is "local," not somehow partitioned, so at the close-in scale it doesn't happen, the two views are virtually identical. This concurrence gives me comfort at the deepest possible emotional level because in the end it does not conflict with the teachings of General Relativity that I believe are substantially correct. What it does do is provide a different interpretation of one aspect of that powerful theory.

END OF PART 2

CHAPTER 16

Critique

What value should any reader place on the GEQ concept as described in the two preceding sections? Should he leap from his chair with barely contained joy at his new understanding of gravity? Or should he throw this book on the floor muttering that here lies yet one more exhibit of the half- baked thinking of the fringe community? I hope for neither extreme, but rather a careful consideration of the ideas with an eye to understanding the strengths and shortcomings of the presentation.

(Before proceeding with my own critique I would be remiss if I did not comment on the objection to the expansion notion put forth in the generally excellent "Math Pages" by an unknown author. He cites a model in which a test particle inside a hollow expanding mass would migrate to the center of the structure as said hollow boundary expands and this is not what happens in reality. His reasoning is wrong because he fails to consider that the expansion is not just relative to one side of the boundary, but the opposite side as well, so in fact its effective location will not change, remaining the same relative distance from all dimensions of inner boundary.)

Now let me start by considering issues in order of their appearance in the text using as a guide, teachings from Newtonian Physics in its range of applicability, those from Special Relativity, and finally those from General Relativity.

It is clear that the GEQ hypothesis cannot disagree with Newton's gravity law and the set of equations describing the expansion process satisfies that requirement except for one statement made about the process. Specifically that "all matter is expanding forever" The question is, "expanding relative to what?"

All three of the reference theories (Newtonian view, SR, and GR) are relativistic in their basics so GEQ must also be relativistic and any glib answer such as *expanding relative to its past size* is unacceptable unless that change can be measured. Results of measurements are the only tool in our kit for distinguishing between truth and fantasy. Without some set of measurements showing such an effect really exists, the entire GEQ concept must be viewed as speculative and/or wrong.

This issue is addressed in chapter 11 on the expansion of the earth where an experiments is described that if carried out could either falsify GEQ if it failed to show accumulative growth of the distance metric over time or support the GEQ hypothesis if it demonstrates that such a change does occur. I cannot use the word *prove* in this latter event for a number of reasons among them, the issue of statistical significance of any results, and the possibility that even if such a result is found, that there is some other explanation I might have missed. Strictly speaking we say you cannot "prove" anything, just show it is likely true to some standard criteria.

Another possible difficulty appearing in the development are the interpretations for equations 6 and 7. On the surface both appear to refer to the size of some representative sphere that goes from its smallest possible size (radius=R_{min}) to infinity, and we simply do not know how to deal with infinite quantities. In chapter 5 about coordinate systems I introduce the idea (dodge) of reversing the expansion view back into the Newtonian form so no infinite size occurs. In this view one is left with a modified Newtonian gravity where the difference between equations 6 and 7 infers a correction in the length metric that enables the calculation of any physical phenomena with slight difference in magnitude compared to a strictly Newtonian view. One example of this approach is the calculation of the so-called anomalous acceleration of Pioneer 10 and 11 (chapter 12).

While this approach is successful for calculations it implies that GEQ is nothing more than a Newtonian gravity theory with an addition of a metric correction. This connection is not viable because Newtonian gravity only is valid for small space while GEQ is presented as a large space description incorporating many views from SR and GR.

So it must be taken that GEQ really is making some sort of comment on GR, and while this may be nothing more than changing the metric of length and denying the possibility that spatial expansion faster than light speed exists, I think it is deeper than that, saying GR is a mathematical description of the way gravity works, but it is not a fundamental description of what gravity is.

One final observation about a difference between the GEQ view and GR occurs in the initial conditions chosen for the velocity of expansion of a sphere at its smallest size. For GEQ that velocity is by necessity zero, while the analogous velocity for GR (the boundary of a black hole) is inward and the velocity is undefined, but certainly not zero. When the GEQ process is reversed into the Newtonian form, this gets rid of infinities but does not resolve the difference in point of view.

Moving on to issues about the second law of thermodynamics (chapter 3) I note that the attractive interpretation of gravity appears to violate that law until one looks at the history of how particles of matter got separated by some prior process. In the case examined it is concluded that some earlier work must have supplied extra energy to the system and this satisfies the requirement for ever increasing entropy.

I think an important message here is it is generally useful to understand that the present state of any system is dependent on earlier actions and when considering how mechanical processes play out, it can be useful to consider the system history. This approach can clarify some processes where either apparent contradictions appear (the so-called twin paradox) or where interpretations are uncertain. Clearly this insight applies beyond GEQ models. Another message is that expansion is a more appropriate mechanism of the universes' behavior than attraction, but attraction tends to be more in tune with local perception.

The introduction of orbiting with no central force discussed in chapter 6 probably causes discomfort to many readers, but I am certain of its validity. The difficulty is a testament to the power of how paradigms control our perceptions and determine what we interpret as reality. If nothing else in this book is found to be valuable this particular insight (if grasped) should satisfy any reader that time spent on this book was worthwhile.

Chapters 9 and 10, advance of the perihelion of Mercury and bending of starlight by the Sun, respectively, are satisfying, and possibly, necessary, but not sufficient to validate GEQ although the simplicity of the solutions argue for the approach. As already pointed out, the outcomes of the experiments suggested in chapter 11 come closer to accomplishing that task.

Just as interesting historical note on the notion of earth expansion: As far back as 1966 Pasqual Jordan, a German physicist, wrote a massive paper on Paul Dirac's idea that gravity had decreased over time, "*The expansion of the Earth. Conclusions from the Dirac gravitation hypothesis*" . that bears a striking similarity to this present work but fails to notice that all important magical difference between equations 6 and 7 that lies at the heart of the GEQ view. If you let your mind wander a bit you will recognize that saying "gravity has decreased over time" is much the same statement as saying, "the metric has increased over time" and ultimately the same as saying there is a difference between equations 6 and 7.

The anomalous acceleration of the deep space probes (chapter12) happens to be the first strangeness I attempted to explain with the GEQ approach and Dr. John D. Anderson of JPL was very helpful in providing me with unpublished information about the launch and instrumentation details.

His willingness to communicate with me was very encouraging. The analysis I present here lacks statistical details and until and unless NASA reevaluates their data using GEQ models its validity cannot be judged.

If Einstein's happiest thought was a man in free fall doesn't feel his own weight, mine appears in chapter 13 and has immense effect on how I view spiral galaxies and distribution of distant galaxies. It is the following:

When you see the radial velocity of masses increase with distance, you are observing the behavior of mass inside a rigid body or fragments of such a mass. that have inherited some component of that velocity.

This factor should be included in analyzing the behavior of systems where radial velocity distributions do not follow strict Newtonian inverse square law. This approach is in contrast to introducing an ad-hoc factor (dark matter or dark energy) or modifications of Newton's law (MOND).

Chapter 14 on Spiral Galaxies uses this observation to obtain an easy solution to the vexing problem without dark matter or MOND and I like the simplicity of the solution.

Chapter 15 basically about the expansion of the universe, departs more from the canonical view than any other part of this book. In rejecting the notion of spatial expansion and replacing it with mass expansion I believe it connects everything written here previously into a whole, but probably is the most difficult part of the overall theme to accept in spite of its agreement with measurement data.

This sums up what I have to say about GEQ but it leaves open a question: If the theory holds water, where does it fit into the body of knowledge about gravity? I'm not even going to try to answer that question other than my speculations stated above about the GEQ view of GR.

CHAPTER 17

Final Comment

In part 2, I provided six examples showing that changing a single concept in physics allows me to address some physical processes in an atypical manner. I reversed our usual interpretation of gravity's effect. Instead of explaining gravity as an attractive force, this reversal describes it as the consequence of everything expanding. It remains an open issue whether this new interpretation of gravity turns out in the long run to be better, closer to "truth," than the earlier one.

Now, I have a question. Given Einstein's early recognition about the perceptual similarity between gravity's behavior and the experience of an accelerated observer in a closed room, why has this topic not become an active area of investigation in the physics community? I think a clue lies partly in the history of how gravity has been viewed since prehistoric times. A fundamental need exists in *Homo sapiens* to answer the question, "Why?" Once a satisfactory reason is obtained, we cling to the answer as though our lives depend on it. And in a way, they do.

I think the one characteristic that distinguishes *Homo sapiens* from all other life forms on earth is neither toolmaking nor intelligence, but the need to ask "Why?" To the best of my knowledge, no other life form has the capability or need for that kind of questioning beyond its immediate environment.

This need has led to the creation of religious institutions, the sciences, and other civil structures. These tools enable our complex societies to function in more or less continuous and somewhat peaceful ways. However, they also lead to confrontations between all groups holding different sets of beliefs.

The need to hold onto certain viewpoints, to resist changes that upset or modify our beliefs, is no less present in the sciences than in other human endeavors. It is perhaps even more threatening for scientists as the subject-matters they deal with are very complex. Personal competence is hard earned. Nonetheless, it is clearly understood by all scientists that one part of their role is to continue to ask "Why?"

Within the sciences there are certain beliefs, better called "truth markers," that guide researchers. New findings cannot conflict with these understandings. Most of the time, but not always, the truth markers concern topics that have been deeply explored. Conclusions have been reached within the context of the discipline such that they are no longer questioned. Sometimes they are historical, so long believed that they unconsciously become part of comfortable certainties. Gravity is one such belief. The view that it "attracts" is buried in antiquity. None of us living today can know the origins of this interpretation with any specificity.

A few years ago, my wife and I were visiting Alaska. We wandered into an art shop in Anchorage owned by a native Athabaskan. Eventually, we bought a watercolor depicting a fisherman holding a spear, seated in a dug-out canoe, wearing the characteristic wooden hat long ago in vogue. As we were leaving, I asked the owner if he happened to know what the word for gravity was in his native language. After reflecting for a moment, he replied that he did not know it, but that his grandmother certainly would. Using a modern extension of the traditional method of oral history (his cellphone), he called his source of knowledge. Alas, she was not home. He told us to return to his store in a few hours and he would provide us with our answer. We did return later.

He had talked to his grandmother and though she claimed no knowledge of such word, suggested it should be called "That which holds us to the earth." Certainly, this is not definitive, but it is suggestive of a reasonable answer that might occur to a parent when asked by her child why he cannot fly: "Because gravity holds you down."

Consider what any child's reaction would be if the answer given was along the lines of "because everything is getting larger and the earth is pushing up against your feet." Now come nightmares and probably visits to the local child psychiatrist! Clearly the expansion notion is not one that would naturally come to mind based on what our perceptions lead us to believe is happening.

It appears that Einstein came close to questioning the concept's source in his attempts to extend his Special Theory of Relativity into his more extensive General Theory (that included gravity). However, he had different fish to fry, and his insight about the similarity between gravity and acceleration was just a bridge to his remarkable new view of the relationship between matter and space. This new view was by any measure a revolution in thinking. We say it introduced a new paradigm in physics.

Paradigm ... what a powerful word! What is a paradigm? It is everything. We are born with certain capabilities and propensities, not quite a blank slate, but with sensory capabilities that enable us to interact with our surroundings and internal wiring that allows us to develop a notion of self. We are guided by interactions with other entities, both active and passive inputs, and thus become who we are.

In some ways the end product is fragile. In other ways it is strong and demands to be whatever it is. All of us start out questioning. It is in our nature. Some of us are easily satisfied by the answers we receive. Others continue to question. From this latter group come our priests and scientists. They are the creators and repositories of our paradigms.

Paradigms are immensely powerful framers that determine how that inner wiring comes to view and understand the world we inhabit. They tell us who we are, what we believe, and how we relate to other members of our tribes. They *define everything* about us. Is it any surprise that we resist attempts to describe our views as wrong or as misconceptions?

Yet, a primary requirement of science is to question our beliefs and advance our understanding of the universe. This results in a great tension between the need to hold on to the familiar and the attractions of revolutionary thinking. There is a great book written by Thomas Kuhn in 1962 titled, *The Structure of Scientific Revolutions.* It explores how new paradigms arise in a manner that is as fresh and accurate 60 odd years later as the day it was written. I recommend it to anyone who is interested in what paradigms are, how they come to be, and what forces cause them to change.

The ideas about gravity presented in this book do not constitute a new paradigm, though it is conceivable that future investigations might bring them to that level. For now, these teachings are largely in agreement with General Relativity (with modified language and a different view of where GR stands in the links of knowledge). GEQ sees the role of space as *passive* in the gravity dance, only *appearing* as an active player. It sees GR as descriptive rather than causal. It sees Alan Guth's notion of inflation as accurate but extending down to the smallest level.

If a more complete study of the expansion theory shows it is on the whole a viable view, that it changes some perspectives on research in modern physics (that go well beyond the range of knowledge addressed in this book), then perhaps it will become a new link in our understanding of the way the universe works.

ESSAY ON TRUTH IN PHYSICS

I include these materials to make clear what limits should be placed on the ideas in this book. Though I present some data suggesting that these theories may be correct, the standards of claiming "truth" are far more demanding than what is presented here. In the end one cannot "prove" anything. An idea can be brought to a place where, with some level of confidence, one can claim no difference between the notion and a well-defined set of criteria (usually involving many repeated measurements). I am required by proper methodology to offer what I have presented here as speculation, not fact.

The "truth", as referenced in this essay, should be understood in the context of "Classical Physics"; the body of knowledge that started to develop about 150 BC, mainly in Greece. Through the dark ages it grew with input from Egypt and the far east, picked up again after the so-called "Enlightenment" and continued to expand through the beginning of the 20th century, ending with Einstein's development of his Special and General Theories of Relativity. It contains all the work of the early astronomers, Newton's discoveries, thermodynamics and more. The most salient feature of "Classical Physics" is that events described and theorized about are connected by cause-and-effect chains. What we call "Modern Physics" started about 1905. The major method of analysis in all periods of this arcane topic is mathematical and its touchstone of validity is the experiment.

Classical Physics can be distinguished from Modern Physics largely by the issue of statistical relationships that impose certain limits to knowable parameters (such as knowing the exact location and energy levels of objects) and, in some notable cases, behavior that looks more like magic than our everyday cause and effect world.

Note that Einstein's theories of relativity are usually grouped with modern physics even though the descriptions he used are causal and not statistical. They might fit better into the classical period. This being said, Einstein's view of much of modern statistical physics was that he didn't like it. He said, "God does not play dice with the universe". Ironically, some of his work in 1905 led directly to modern physics (with its statistical view of events) and a Nobel Award for Einstein. How about that?

I remember as a young child, perhaps five or six, sitting in a train in Boston's South Station with another train on one side and an empty platform on the other, all at rest. Without warning, the train to my left started to move. For a moment I could not tell if it was moving or if my train had smoothly started down the track. I felt disoriented and it was only when I glanced to my right, seeing that I remained at rest relative to the platform, that I indeed knew that the train to my left was in motion (in a well-defined setting including me "the observer", the observed train and the platform).

The point of this is that the "truth" of what is happening depends on perceptions of events and all relevant circumstances. This is very different from simply saying "such and such is happening" based on a kind of certainty that immediate perception is the complete truth, an assumed property of some object. The actual event described is only loosely specified.

This is just one example that hints at the difference between "truth" in physics and "truth in everyday thinking". Consider the following: In mathematics I can clearly define a straight line as the shortest path between two points drawn on a flat piece of paper or, perhaps more appropriately, two points floating in empty space. This is a kind of abstraction that satisfies us and surely any two individuals will agree with the definition.

However, in physics that agreement is conditional and requires more information about the states of any observers "relative" to the observed and for that matter, what is meant by "flat paper" or "empty space.

If, for example, I ask an observer who is "at rest" relative to two points in space, he will report a line drawn between them is a straight line (see fig. below left). If I have a second observer moving at constant velocity towards the two separated two points and I ask him the same question, he will report the line is not left to right but slanted at some angle (see fig. below center). More dramatically, if his motion towards the two points is increasing in time, if he is accelerating, he will report the line drawn is a curve (see fig. below right).

PROPERTY OF MOVING BODIES AS FUNCTION OF OBSERVER'S MOTION

These are simple examples and it is easy to understand how other consid-erations can cloud the clarity of what is "true". Take for example the dynamic relationship between the sun and earth. While we say with confidence "earth circles the sun", earlier opinions were reversed to "the sun circles the earth." In fact, considering this issue from a purely logical point of view (ignoring all the other planets), either interpretation is valid. You can convince your-self of this with a simple exercise using paper and pencil. By drawing both versions either can be shown to be possible.

But now consider how other beliefs might affect your conclusion: In particular imagine some authority figure, perhaps the leader of your religious group, who states the opinion that mankind is the most important factor in existence. Therefore, it is clearly true that he is the center of the universe and it is only appropriate that the sun circles the earth.

It is easy to see how this opinion, based on long held beliefs, can cause this notion to be widely believed. Even if other information about the heavens is slowly accrued by a few individuals, the earlier belief, held by many people for a long time and stated as a "truth", can overwhelm this "new" belief held by just a few.

Beyond that, one can easily imagine the two different views causing a schism, erupting into disputes that ultimately, probably at first, cause the new view to be suppressed in not gentle ways. Individuals holding these new ideas could conceivably be branded as heretics and, one way or another, punished for their beliefs. This is just a simple example of a rich and potentially complicated set of events. One imagines, quite correctly, that out of these conflicts arise groupings or institutions that work with great diligence to define "reality" in accordance with their set of beliefs.

It is in the nature of all nature, whether one refers to mankind or other processes, that each will tend to act in such a manner as to preserve its own existence and belief systems. So one understands, or should understand, that "truth" is a slippery fellow that for the would-be scientist requires some extra vigilance and careful methods to gain any hope of approaching what today we, in the west at least, call verifiable repeatable "facts".

Even then, assuming some successful progress towards this goal, understand these "truths" are always "relative to the observer" and not absolute in any sense. Given that we as a species have limited existence, both in the sense of time and space, our "truths" are limited likewise and should not be held onto with excess conviction. The lesson is "Hold Gently" and be prepared at any time to let go as new information is uncovered.

In order to penetrate these difficulties science has slowly evolved "The Scientific Method", a series of more or less well-defined steps to determine what is considered "true" and what is considered either outright "false" or "speculative". In the latter case, more study and evidence are required. I will have a lot more to say about these steps later.

In the end it is information, gathered through all our senses, that makes up our body of knowledge that has evolved into what we refer to as "science". Given the limits of our perceptions and the (sometimes) disputation of generally held beliefs, it is not surprising that scientific "truths" have followed a jiggered course with many modifications over time. Sometimes science moves in the right direction (according to later judgements) and sometimes it veers in the wrong direction as more information becomes available.

There is an interesting principle called "Occam's Razor", not really part of the so-called "Scientific Method", that is often used to distinguish among different explanations of some phenomenon, where all explanations, according to the methods of analysis, appear to be plausible. According to Occam's principle, the simplest explanation should be chosen. Of course one does not simply throw away the more complex solutions, but rather studies them thoroughly. Hopefully, clarity emerges.

At this point I would like to give you an example of this kind of situation which caused great pain to Galileo and took many years to untangle

Just before Christ's time Ptolemy, an Egyptian astronomer of Greek descent, described the solar system with earth at the center and the sun circling the earth. All the other planets were likewise arrayed in circular orbits around the earth. In essence this model agreed with the earlier Greek belief (read Aristotle) that the circle was a sign of perfection and hence must reflect that which is required for nature. At the time enough measurements had shown that the planets did not travel in circles. Instead their direction of motion as seen from the earth reversed in an orderly manner: This is referred to as "rétrograde motion". To account for this, Ptolemy claimed that each planet travelled in a small circle along the track of a larger circle that traced out its passage around the earth.

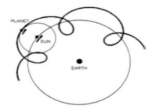

RETROGRADE MOTION

While this scheme seems most extraordinary from our point of view, it did explain what was measured from the earth's location. More importantly, from the point of view of the Catholic Church, soon to become dominant in Europe, at perhaps the beginning of what we now call the "Dark Ages", the earth centric idea won out and little occurred to change this until the time of Copernicus some 1500 years later.

Copernicus, as were many astronomers, was a member of the church (and a Polish nobleman besides). He made his celestial observations from the roof of his monastery. Eventually he came up with the idea that the sun was at the center of the solar system and wrote a book titled, *"On the Revolutions of the Celestial Spheres"*. He had the good sense to get dead before the book was published. The church could do nothing to him but added the book to the list of forbidden texts. There it remained until 1835.

Around 1600 came Kepler (German) and Galileo (Italian) both of whom adopted Copernicus's idea that the sun was at the center, not the earth. They also got rid of the notion that all orbits were perfect circles and hence explained the appearance of retrograde motions without Ptolemy's epicycles. Certainly, Kepler's ideas were not beloved by the church, but unlike Galileo, the full disapproval of that by now powerful institution did not come down on his head. The reformation of the German church had occurred some 100 years previously. As a Lutheran, he was outside the Catholic Church's range of retribution. Not so Galileo.

In 1610 Galileo published a book titled, "*Starry Messenger*", putting forth his ideas about a heliocentric organization of the solar system. The church pounced. What followed were two responses; the first in 1616 declaring his ideas heretical and that he should neither teach nor believe in those ideas. Evidently Galileo either didn't get the message or thought he could get away with a bit of misdirection so in 1633 he wrote a second book comparing Copernicus's idea of a heliocentric solar system to one with the earth at the center. The church didn't buy his claim that in the book he didn't choose between them and awarded him with a second trial placing him under house arrest for the remainder of his life. He died nine years later in 1642.

So much for putting forth unpopular ideas! The point here is that if Occam's principle had been applied to the comparative beliefs the far simpler one, a heliocentric solar system, would have been selected.

Now where do these ideas come from that ultimately *are* interpreted as what we call truths? They come from a variety of sources. Some are the results of measurements of ever-increasing accuracy. Others from "what if" thoughts in the minds of investigators. They come from mathematical analyses of models purporting to be descriptions of dynamic physical systems. In the end, one can only say that they come in a variety of ways and for a variety reasons, but they do come by the boatload.

The issue is: How do these ideas get sorted out, verified, or rejected? As previously mentioned, one major tool is the so-called "Scientific Method".

The Scientific Method consists of a series of steps. First, the idea is phrased as a question (why does thus and such occur?) or as a "what-if" statement (what if thus and such is true?). What follows generally is a statement or "model" of the idea which can be analyzed by some means, usually mathematics. Next comes the testing of the idea by one or more experiments. Often the results of the experiment modify the model to some degree. Finally, the model and the experimental results are published, presented for further analysis by the scientific community. Ideas that fit into the prevailing view, only filling in details, are more likely to be accepted than ideas that lay out an entirely new paradigm or parsing of an issue. In either case the community as a whole examines all aspects of the new notion, often designing and carrying out additional experiments to test it. Eventually the scientific community accepts or refutes the idea. If the theory is accepted it becomes a new link in the chain of understanding. Additional papers on the topic appear in the literature of the field.

It is especially important to note that the golden touchstone of "truth" in science is *measurement*, not theory. Following any measurement results, validating, or invalidating some idea, the scientific method requires additional measurements by other members of the community to assure that the results are repeatable. In most cases the measurements are repeated many times under the same conditions or variations that in principle should give equivalent results. The basic idea in repeating the experiments is to assure that biases do not cloud the issue.

Most of the time data collected in this way does not provide exact results, but rather a scattering. This requires careful statistical analysis in order that some level of confidence can be attributed to the results. Hence, even after many repeated measurements, the notion of "truth" is graded according to well established criteria used to compare a data set to a theoretical expectation. If the mean value of a data set meets certain criteria, one can say, "*the null hypothesis(that there is no difference between the theoretical mean and the calculated mean) is accepted with a strong possibility.* Note, it is this *comparison* with some external data that is meaningful. If no such external reference data exists, one cannot claim the result represents a "truth", only that the topic might be worthy of further study. The results might later become part of the chain of evidence. I will elaborate on this last point a bit later.

The statistical methods used are well defined. They include criteria for deciding with what level of confidence the community should regard any new idea (*after* sufficient tests of the idea have been made by different members of the community). Eventually, a *confidence level* is established. What this interval means is that if the reference value, the "truth" value, for the phenomenon being studied lies within the desired confidence interval, the Null Hypothesis has not been refuted. Notice, even then the new idea has not been proved, but *the findings have not shown any difference from the theory's predictions.*

The confidence level placed on any data set varies with discipline: In the social sciences generally a (95%) level is considered significant. In quantum physics a (99.9994%) level is required.

I previously mentioned a situation in which a new theory is being investigated, probably in the context of trying to explain some phenomenon not previously explained by existing theory. In that case there is no "truth" to compare the results to. The validity of the theory, in accordance with the scientific method, must be considered a conjecture as opposed to a truth. What then?

The theory may be a starting point in a chain of evidence, but more is required. The new theory must be examined in detail by the community to see if its claims are logically and mathematically viable, making sure it does not conflict with other "truths" possibly missed in theorizing. If it passes all these queries, testing must begin by other members of the community. Finally, if these additional efforts result in no disagreements with the theory, just maybe one can claim a discovery with a degree of confidence, as in the range of evaluation discussed above.

Note that this issue is particularly relevant to the goal of this book as I have presented a modified interpretation of Newton's gravity law. While I have also present some supportive data, in essence it is insufficient to validate my claims. In view of this and the above considerations I understand clearly that my theory is speculative, requiring extensive examination by the physics community. I make no claims beyond that as to the value of this work. However, as you can see in chapter 11, I describe an experiment that can be used to test the main hypothesis. If that test is carried out it will provide one more link to help decide if my ideas might be valid or should be rejected. And so it goes, with very great care, physics will weed out ideas that seem good but fail the stringent tests of experiments.

From this you can see that scientific communities place extraordinarily strong constraints on themselves and generally one should accept their claims.

On the other hand, while from the point of view of "truth" it might be desirable to say that this scientific approach insures against squabbles, biases, fights, backbiting and nastiness in all its various forms that infiltrate most human activities. This is of course not the case in reality. Scientists are in the end human; the meetings and writings of their journals are filled with the same intemperance that occurs elsewhere in human endeavors. In the end this is good. It makes science an exciting and lively pursuit to follow with enough emotional food for all types of personalities.

I think it is worthwhile at this point to give an example of how the scientific method clarified the community's belief about how light moved through empty space. The prevailing belief was that the way sound propagates was an analogons model. All space was filled with "luminescent aether" that played the same role for light as air did for sound.

In approximately 1680 Newton was among the first to investigate the speed of sound. He got it wrong, but a few years later a French scientist, Pierre-Simon Laplace, corrected his error and derived the correct equation. The mechanical model they used at the time is more or less equivalent to our present view. The important point is that sound can be described as a longitudinal wave causing pressure in a medium, generally air, where its velocity is determined by the air's density.

There was a serious problem with the aether idea in that the velocity of light was so fast that the density of the aether was extremely high. It was unclear as to how physical bodies such as the earth could move through it unencumbered. Thus, the theory had to include the hypothesis that only light energy was affected by its presence.

In 1881 an American physicist, Albert A. Michelson, designed a device (an interferometer) that could measure the speed of light to a high degree of precision. He reasoned that since the earth was traveling through space at about 30 kilometers per second in its orbit around the sun, he could use his device to measure light's speed in different directions, and thus detect the "aether wind" as it was called. His model was as indicated below. Since the speed of the earth is only about .01% of that of light, he knew the difference he would have to measure was exceedingly small indeed.

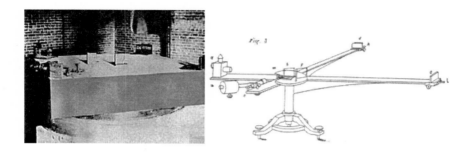

MICHELSON-MORLEY INTERFEROMETER

His measurements showed no significant difference in any direction. Later analysis, by himself and others, showed his experimental error, variations in multiple tries, due to limits from his device capabilities, were larger than the expected difference in velocity from the theory. This is the careful analysis cited as a requirement in the Scientific Method.

In 1885, Michelson combined forces with another American, Edward W. Morley, a chemist. The two of them refined Michelson's interferometer, corrected some other issues with the experiment and repeated the measurements. Again, no significant differences were measured! What is of particular interest is that they believed the aether theory was correct. The point of the second series of measurements done in 1887 was specifically to show that the aether did exist! Nowadays these experiments are often referred to as the most fortuitous, failed results in classical physics! The explanation for their "failed findings" led to what are now called the Lorentz contraction equations and a bit later (1905) Einstein's Special Theory of Relativity.

The fact is that the scientific community did not give up on the aether idea all that easily and even today some scientists still think it is a real possibility. One thought was that the earth dragged the aether along with it, but Sir Oliver Lodge performed an experiment showing this was not true. Also, the Lorentz contraction (objects like meter sticks get shorter in the direction of motion) has been cited as a reason for its failure, but the success of Einstein's Special Theory of Relativity and its more complete theory, including gravity (1915), has pretty much put the controversy to rest. In 1905 Albert Einstein published his special theory of relativity, which implied that there was no need for the mysterious and undetectable aether. All of the current experimental contradictions could be explained away by the new theory. The aether could be forgotten.

As the years have passed many experiments in classical physics have resulted in explanations of certain phenomena and theories that are taken to be unquestionably "true". These theories serve as signposts to guide and constrain new ideas. Any new ideas or new interpretations of old ones cannot disagree with facts known to be true in the sense that the physics community has given its blessings to these repeatedly measured notions and found them to be invariant and accurate in their predictions. This is not to say more information gained does not limit their range of applicability, one notable example being the relationship between Newton's gravity law and Einstein's General Theory of Relativity (GR). GR reduces to Newton's Law when velocities are small compared to light velocity and distances are not very large. While Newton's theory is correct, it is a description of gravity's behavior constrained within certain limits.

There were at least two statements in Newton's law for gravity that were questionable and Einstein corrected them. The first was that the effects of gravity were instantaneous. According to Newton if the sun suddenly vanished, the effect would be immediately felt here on earth. In Newton's time the notion of "field" was not yet known and while Newton was uncomfortable with this result (he commented about it in a letter) it remained a part of his theory. It was labeled "action at a distance".

The fields concept was well established by the time Einstein came on the scene (1905) and Einstein pointed out that causal action could not exceed the speed of light. The second correction was also about time. Combined with the first issue, it proved highly consequential in how Einstein developed his Special Theory of Relativity. Newton imagined "time" was the same everywhere, as though there existed somewhere out in space a great timepiece that coordinated time for everything (regardless of states of motion or locations). Einstein's notion that the velocity of light is a constant for any and all observers meant that this could not be true. The time of any event must be modified to account for the views of different observers (depending on their relative distances and states of motion).

For example, if two observers, one on a rapidly moving train and the other standing beside the track (as shown below), see a light located in the center of a railroad car abruptly turn on just as the car passes the observer on the ground, the observer on the train will see the light beam reach the front and back of the car *simultaneously.* The observer on the ground will report the light beam hit the back of the car first, then the front of the car, *not simultaneously.*

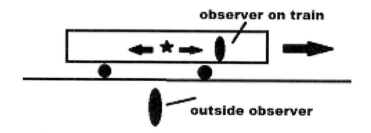

NO MEANING TO SIMULTANEITY

This is because while both he and the observer on the train see light velocity the same, the observer on the ground sees the back of the train move towards the light source and the front of the train move away from it. Even more dramatically, if two identical twins separate, one staying on earth and the other taking a very fast ride on a rocket-ship, the one who took the fast ride will be younger than the one who stayed at home when they reunite!

Because of these time parameter differences between the views of Newton and Einstein, Einstein introduced the dimension of "time" in his description of space, what is now called "space-time". While Newton used what is called 3-dimensional *Euclidian Space,* what we all studied in high-school geometry, Einstein used 4-dimensional space-time, called *Minkowski Geometry.*

In spite of these corrections, Newton's gravity law is one of those beacons of truth that remain as true today as it did in the 17th century. Subject to the limitations discussed above, no new idea in physics can fail to agree with its bones if the idea is to be considered possibly real. However, what can be allowed is re-interpretation of Newton's gravity law or its use in the form of approximate solutions for Einstein's much more complex field equations (developed in his General Theory of Relativity). This latter interpretation operates under the rubric of "Post Newtonian Physics". One well known example is the distance light travels in empty space (called a geodesic). Light does not travel in a straight line, but in a way analogous to the shortest flight path of an aircraft on earth between two distant locations. The path curves along the circumference of the earth.

The 2nd Law of Thermodynamics is one additional truth marker from classical physics that plays a central role in this book. Up until roughly 1640 the physics community believed that "heat" was a fluid. Between 1640 and 1650 a more complete analysis showed that it was another form of energy and an exactly accurate description of its behavior was developed. There are three Laws and it is this 2nd law that is of prime importance for examining the behavior of Newton's gravity law.

The 2nd law indicates the irreversibility of natural processes and the tendency of natural processes to lead towards spatial homogeneity of matter and energy. It can be formulated in a variety of interesting and important ways. It implies the existence of a quantity called the entropy of a system. Entropy always increases towards disorder, and in terms of any system, this means less energy per unit volume is available. One example is the cooling of a heated object. As heat energy flows to the cooler surroundings, less of a temperature gradient is available for any heat engine. Every time you experience a hot cup of coffee cool off while it sits on the table in front of you, you are encountering this action.

The 2nd law of thermodynamics was eventually understood to apply to all forms of energy, not just heat. If any known law of physics can be said to never be violated in the long run, surely the 2nd Law is one of them. A broad definition is that any high energy region of space will tend to spread its energy to its lower energy surroundings. It is this definition that is of particular relevance to Newton's law in terms of "energy dispersal". To quote, "Increase of entropy in a thermodynamic process can be described in terms of "energy dispersal" and the *spreading of energy*," while avoiding all mention of "disorder" and "chaos"---."

Another way of understanding the effect of ever-increasing entropy is that there are a lot many more ways to be wrong about why some process behaves as it does, and the more observers involved only increases the uncertainty unless they are all using the same methods of analysis. The natural tendency without proper training is chaotic and not to be trusted.

It is in the nature of nature that a consistent understanding of what is going on tends to go from hard earned clarity to fuzzy uncertainty as more and more points of view are brought into the discussion. From all of this, understand that having an idea about the way some process works does not count as a viable theory until it is subjected to testing by the community. Be cautious in accepting new untested claims. Based on the the way of entropy, it is more likely to be wrong than right.